T0269209

CAMBRIDGE LIBRARY COLLECTION

Books of enduring scholarly value

Physical Sciences

From ancient times, humans have tried to understand the workings of the world around them. The roots of modern physical science go back to the very earliest mechanical devices such as levers and rollers, the mixing of paints and dyes, and the importance of the heavenly bodies in early religious observance and navigation. The physical sciences as we know them today began to emerge as independent academic subjects during the early modern period, in the work of Newton and other 'natural philosophers', and numerous sub-disciplines developed during the centuries that followed. This part of the Cambridge Library Collection is devoted to landmark publications in this area which will be of interest to historians of science concerned with individual scientists, particular discoveries, and advances in scientific method, or with the establishment and development of scientific institutions around the world.

The Life of Sir Humphry Davy

Sir Humphry Davy (1778–1829) was a hugely influential chemist, inventor, and public lecturer who is recognised as one of the first professional scientists. He was apprenticed to an apothecary in 1795, which formed his introduction to chemical experiments. A chance meeting with Davis Giddy in 1798 introduced Davy into the wider scientific community, and in 1800 he was invited to a post at the Royal Institution, where he lectured to great acclaim. These volumes, first published in 1831, contain Davy's official biography. Researched and written by John Ayrton Paris, the work describes in detail Davy's life and his scientific studies. Organised chronologically with excerpts from his private correspondence, Davy's early life and his experiments and lectures at the Royal Institution and his Presidency of the Royal Society between 1820 and 1827 are explored in vivid detail. Volume 1 describes his life and work until 1812.

Cambridge University Press has long been a pioneer in the reissuing of out-of-print titles from its own backlist, producing digital reprints of books that are still sought after by scholars and students but could not be reprinted economically using traditional technology. The Cambridge Library Collection extends this activity to a wider range of books which are still of importance to researchers and professionals, either for the source material they contain, or as landmarks in the history of their academic discipline.

Drawing from the world-renowned collections in the Cambridge University Library, and guided by the advice of experts in each subject area, Cambridge University Press is using state-of-the-art scanning machines in its own Printing House to capture the content of each book selected for inclusion. The files are processed to give a consistently clear, crisp image, and the books finished to the high quality standard for which the Press is recognised around the world. The latest print-on-demand technology ensures that the books will remain available indefinitely, and that orders for single or multiple copies can quickly be supplied.

The Cambridge Library Collection will bring back to life books of enduring scholarly value (including out-of-copyright works originally issued by other publishers) across a wide range of disciplines in the humanities and social sciences and in science and technology.

The Life of Sir Humphry Davy

Volume 1

John Ayrton Paris

CAMBRIDGE UNIVERSITY PRESS

Cambridge, New York, Melbourne, Madrid, Cape Town,
Singapore, São Paolo, Delhi, Tokyo, Mexico City

Published in the United States of America by Cambridge University Press, New York

www.cambridge.org
Information on this title: www.cambridge.org/9781108073189

This edition first published 1831
This digitally printed version 2011

ISBN 978-1-108-07318-9 Paperback

Sir Thos. Lawrence pinxt. W. H. Worthington sculpt.

H Davy

Engraved by permission of the Council of the Royal Society from the original painting.

London. Published by Henry Colburn & Richard Bentley. 1831.

THE LIFE

OF

SIR HUMPHRY DAVY,

BART. LL.D.

LATE PRESIDENT OF THE ROYAL SOCIETY, FOREIGN ASSOCIATE
OF THE ROYAL INSTITUTE OF FRANCE,
&c. &c. &c.

BY

JOHN AYRTON PARIS, M.D. CANTAB. F.R.S. &c.
FELLOW OF THE ROYAL COLLEGE OF PHYSICIANS.

IN TWO VOLUMES.

VOL. I.

LONDON:
HENRY COLBURN AND RICHARD BENTLEY,
NEW BURLINGTON STREET.
M DCCC XXXI.

TO

HIS ROYAL HIGHNESS

PRINCE AUGUSTUS FREDERICK,

DUKE OF SUSSEX, K.G. D.C.L.

&c. &c. &c.

PRESIDENT OF THE ROYAL SOCIETY;

THESE MEMOIRS OF A PHILOSOPHER
WHOSE SPLENDID DISCOVERIES
ILLUMINED THE AGE IN WHICH HE LIVED,
ADORNED THE COUNTRY WHICH GAVE HIM BIRTH,
AND OBTAINED FROM FOREIGN AND HOSTILE NATIONS
THE HOMAGE OF ADMIRATION AND
THE MEED OF GRATITUDE,
ARE,
BY THE GRACIOUS PERMISSION OF HIS ROYAL HIGHNESS,
DEDICATED WITH SENTIMENTS OF PROFOUND RESPECT,

BY

THE AUTHOR.

PREFACE.

THE reflecting portion of mankind has ever felt desirous of becoming acquainted with the origin, progress, habits, and peculiarities of those whom the powers of genius may have raised above the plane of intellectual equality; but neither the nature of the information, nor the extent of the detail that may be necessary to satisfy so laudable a curiosity, can ever be estimated by any common standard, since it is not in our nature to contemplate an object of admiration, but with reference to our own predilections and sympathies; and hence every reader will form a scale for himself, according to the degree of interest he may feel for the particular character under review. The Poetical enthusiast, who could not sufficiently express his gratitude on being told that Milton wore shoe-buckles, would very probably not have given 'four farthings,' as Gray says, to know that the shoes of Davy were tanned by catechu; and yet if the relative value of this information were fairly estimated, it must be admitted that the former is a matter of barren curiosity, the latter, a fact of some practical utility. In a word, we very naturally connect the man with his works, and we care not to extend our acquaintance with the one, but in proportion as we have derived pleasure from the other.

In like manner, very different estimates will be formed of the degree of praise due to a distinguished philosopher, because the few who are deeply imbued with a knowledge of the science he may have adorned and enlightened must not only appreciate the value of his labours, but understand the difficulties which opposed the accomplishment of them, before they can arrive at a sound decision upon the question: and here again the judgment of the most scientific may be unfortunately warped; it may be corrupted by secret passions or sinister influences; be distorted by the prejudices of education and habit, or unduly biassed by invincible prepossessions.

No man ever soared, like an eagle, to the pinnacle of fame, without exciting the envy and perhaps the hatred of those who could only crawl up half-way; while, on the other hand, where no rivalry can exist, the splendour of such an ascent will captivate the bystander, and by exciting intemperate triumph and unqualified admiration, change without diminishing the sources of erroneous judgment, and substitute adulation for calumny. Under such circumstances, an allusion even to the common frailties of genius becomes offensive; the biographer is called upon for the delineation of a perfect man; but the world is satisfied with nothing short of 'a faultless monster;' and yet, while they would impose upon him the same restraint as Queen Elizabeth laid upon her artist — to execute a portrait without a single shadow, they little imagine how completely they obscure the features of their idol, by the haze of incense in which they continually envelope it. These are evils against which a future historian will not have to contend; for time tries the characters of men, as the furnace assays the quality of metals, by disengaging the impurities, dissipating the superficial irridescence, and leaving the sterling gold bright and pure.

Nor can the extent of our obligations to a philosopher be appreciated until time shall have shown the various important purposes to which his discoveries may administer. The names of Mayow and Hales might have been lost in the stream of discovery, had not the results of Priestley and Lavoisier shown the value and importance of their statical experiments on the chemical relations of air to other substances. The discoveries of Dr. Black on the subject of *latent* heat could never have obtained that celebrity they now enjoy, had not Mr. Watt availed himself of their application for the improvement of the steam-engine; and the views of Sir H. Davy respecting the true nature of chlorine become daily more important from the discovery of new elements of an analogous nature. In future ages, the metals of the alkalies and earths may admit of applications, and open new avenues of knowledge, of which at present we can form no idea; but it is obvious that, in the page of history, his name will gather fame in proportion as such discoveries unfold themselves.

It must be admitted, that such considerations may furnish an argument against the propriety of writing the life of a contemporaneous philosopher; and yet I will never admit, with Mr. Babbage, that "the volume of his biography should be sealed, until the warm feelings of surviving kindred and admiring friends shall be cold as the grave, from which remembrance vainly recalls his cherished form, invested with all the life and energy of recent existence."

Is it not possible that the errors of partiality, which have so frequently been charged upon the writer on these occasions, may often be ascribed, with greater truth and justice, to the prejudices of the reader—that, after all, the distortion might not have existed in the portrait itself, but in the optics of the observer? Such an opinion, however, even were it true, carries along with it no consolation to the biographer; for I know of no method by which the picture can be adapted to the focus of every eye.

If, however, contemporaneous biography has its difficulties and impediments, so has it also its advantages. Dr. Johnson has remarked, in his Life of Addison, that "History may be formed from permanent monuments and records; but Lives can only be written from personal knowledge, which is growing every day less, and in a short time is lost for ever. The delicate features of the mind, the nice discriminations of character, and the minute peculiarities of conduct, are soon obliterated."

I did not enter upon this arduous and delicate task, without a distinct conception of the various difficulties which would necessarily oppose its accomplishment. I well knew that the biographer of Davy must hold himself prepared for the dissatisfaction of one party at the commendations he might bestow, and for the displeasure of the other at the penury of his praise, or the asperity of his criticism.

After great labour and much anxiety, I have at length completed the work; and in giving it to the world, I shall apply to myself the words of Swift—"I have the ambition to wish, at least, that both parties may think me in the right; but if that is not to be hoped for, my next wish should be, that both might think me in the wrong, which I would understand as an ample justification of myself, and a sure ground to believe that I have proceeded, at least, with impartiality, and perhaps even with truth."

It is certainly due to myself, and perhaps to the world, to state the circumstances by which I was induced to undertake a work requiring for its completion a freedom from anxiety, and an extent of research, scarcely compatible with the occupations of a laborious profession; and which, I may add, has been wholly composed during night, in hours stolen from sleep. Very shortly after the death of Sir

Humphry Davy, an account of his life, written by no friendly hand, nor 'honest chronicler,' was submitted for my judgment by a Journalist who had intended to insert it in his paper. At my request, it was committed to the flames; but not until I had promised to supply the loss by another memoir. The sketches by which I redeemed this pledge were published in a weekly journal—THE SPECTATOR; and they have since been copied into various other works, sometimes with, but frequently without any acknowledgment. They constitute the greater part of the Life which was printed in the Annual Obituary for 1829; and they form the introduction to an edition of his "Last Days," lately published in America.

I was soon recognised as the writer of these Sketches; and the leading publishers of the day urged me to undertake a more extended work. To these solicitations I returned a direct refusal: I even declined entering upon any conversation on the occasion; feeling that the wishes of Lady Davy, at that time on the Continent, ought in the very first instance to be consulted on the subject. Had not the common courtesy of society required such a mark of attention, the wish expressed by Sir Humphry in his Will would have rendered it an imperative duty. On her arrival in London, in consequence of a letter she had addressed to Mr. Murray, I requested an interview with her Ladyship, from whom I received not only an unqualified permission to become the biographer of her illustrious husband, but also the several documents which are published with acknowledgement in these memoirs. I still felt that Dr. Davy might desire to accomplish the task of recording the scientific services of his distinguished brother; and, had that been the case, I should most undoubtedly have retired without the least hesitation or reluctance; but I was assured by those who were best calculated to form an opinion upon this point,—for he was himself absent from England,—that motives of delicacy which it was easy to appreciate, would at once lead him to decline an undertaking embarrassed with so many personal considerations. The task, however, of collating the various works of Sir H. Davy, and of enriching them by notes derived from his own knowledge of the circumstances under which they were written, I do hope will be accomplished by one who is so well calculated to heighten the interest, and to increase the value of labours of such infinite importance to science, and to the best interests of mankind.

The engraving which adorns the volume is from a painting by Sir Thomas Lawrence, presented to the Royal Society by Lady Davy; and I beg the Council of that learned body to accept my thanks for the permission they so readily granted for its being engraved. It is one of the happiest efforts of the distinguished Artist, and is the only portrait I have seen in which his features are happily animated with the expression of the poet, and whose eye is bent to pursue the flights of his imagination through unexplored regions.

I must also embrace this opportunity of publicly expressing my thanks to the Managers of the Royal Institution, who, in the most handsome manner, immediately complied with my request to inspect their Journals, and to make such extracts from them, as I might consider necessary for the completion of my memoirs.

To Mr. Davies Gilbert, I am under obligations which it is difficult for me to acknowledge in adequate terms, not only for the value of the materials with which he has furnished me, but for the kindness and urbanity with which they were communicated, and for the ready and powerful assistance which I have so constantly received from him during the progress of the work.

To the other enlightened individuals from whom I have received support, I have acknowledged my obligations in the body of the work; and should I have inadvertently passed over any service without a becoming notice, I trust the extent of the labour and the circumstances under which it has been performed, will plead my apology.

Dover Street, January 1, 1831.

CONTENTS.

CHAPTER I.

CHAPTER II.

CHAPTER III.

CHAPTER IV.

CHAPTER V.

CHAPTER VI.

CHAPTER VII.

CHAPTER VIII.

CHAPTER IX.

Octr. 19th

When Potash was introduced into a tube having a platina wire attached to it so & fused into the tube so as to be a conductor ie so as to contain just water enough slightly solid —

— & inserted over. mercury, when the Platina was made the neg —

No gas was formed & the

mercury became oxydated —

& a small quantity of the

alkaligen was produced round

the plat: wire. as was evident

from its [illegible] inflammation

by the action of water

— When the mercury was

No gas was formed & the

mercury became oxydated —

& a small quantity of the

alkaligen was produced round

the plat: wire, as was evident

from its effects: inflammation

by the action of water

— When the mercury was

made the neg: gas was develo[ped]
in great quantities.. from the
pos: wire ~~[illegible]~~ gas & oxide
from the neg mercury &
this gas proved to be pure
oxygene by Capt: Exp^t.

proving the decomp^n
of Potash

THE LIFE

OF

SIR HUMPHRY DAVY,

BART. &c. &c.

CHAPTER I.

Birth and family of Sir H. Davy.—Davy placed at a preparatory school.—His peculiarities when a boy.—Anecdotes.—He is admitted into the grammar-school at Penzance.—Finishes his education under Dr. Cardew at Truro.—Death of his father.—He is apprenticed by his mother to Mr. John Bingham Borlase, a surgeon and apothecary.—He enters upon the study of Chemistry, and devotes more time to Philosophy than to Physic.—The influence of early impressions illustrated.—His poetical talent.—Specimens of his versification.—An Epic Poem composed by him at the age of twelve years.—His first original experiment in chemistry.—He conceives a new theory of heat and light.—His ingenious experiment to demonstrate its truth.—He becomes known to Mr. Davies Gilbert, the founder of his future fortunes.—Mr. Gregory Watt arrives at Penzance, and lodges in the house of Mrs. Davy.—The visit of Dr. Beddoes and Professor Hailstone to Cornwall.—The correspondence between Dr. Beddoes and Mr. Davies Gilbert, relative to the Pneumatic Institution at Bristol, and the proposed appointment of Davy.—His final departure from his native town.

HUMPHRY DAVY was born at Penzance, in Cornwall, on the 17th of December 1778.* His ances-

* I have been favoured by the Rev. C. Val. Le Grice, of Trereiffe, with the following extract from the Parish Register, kept at Ma-

tors had long possessed a small estate at Varfell, in the parish of Ludgvan, in the Mount's Bay, on which they resided: this appears from tablets in the church, one of which bears a date as far back as 1635. We are, however, unable to ascend higher in the pedigree than to his paternal grandfather, who seems to have been a builder of considerable repute in the west of Cornwall, and is said to have planned and erected the mansion of *Trelissick*, near Truro, at present the property and residence of Thomas Daniel, Esq.

His son, the parent of the illustrious subject of our history, was sent to London, and apprenticed to a carver in wood, but, on the death of his father, who, although originally a younger son, had latterly become the representative of the family, he found himself in the possession of a patrimony amply competent for the supply of his limited desires, and therefore pursued his art rather as an object of amusement than one of necessity: in the town and neighbourhood of Penzance, however, there remain many specimens of his skill; and I have myself seen several chimney-pieces curiously embellished by his chisel.*

dron:—"Humphry Davy, son of Robert Davy, baptized at Penzance, January 22, 1779." The house in which he was born has been pulled down and lately rebuilt.

* Soon after the days of Gibbons, the art of ornamental carving in wood began to decay, and it may now be considered as nearly lost. Its decline may be attributed to two causes. In the first place, to the change of taste in fitting up the interior of our mansions; and in the next, to the introduction of composition for the enrichment of picture-frames and other objects of ornament. "Robert Davy," says a correspondent, "has been considered in

I am not able to discover that he was remarkable for any peculiarity of intellect; he passed through life without bustle, and quitted it with the usual regrets of friends and relatives. The habits, however, generally imputed to him were certainly not such as would have induced us to anticipate a high degree of steadiness in the son.

His wife, whose maiden name was Grace Millett, was remarkable for the placidity of her temper, and for the amiable and benevolent tendency of her disposition: she had been adopted and brought up, together with her two sisters, under circumstances of affecting interest, by Mr. John Tonkin, an eminent surgeon and apothecary in Penzance; a person of very considerable natural endowments, and whose Socratic sayings are, to this day, proverbial with many of the older inhabitants.

To withhold a narrative of the circumstances that led Mr. Tonkin to the adoption of these orphan children, would be a species of historical fraud and literary injustice, by which the world would not only lose one of those bright examples of pure and disinterested benevolence, which cheer the heart and ornament our nature, but the medical profession would be deprived of an additional claim to that public veneration and regard, to which the kind sympathy of its professors has so universally entitled it.

The parents of these children, having been attacked by a fatal fever, expired within a few hours

this neighbourhood as the LAST OF THE CARVERS, and from his small size, was generally called *The little Carver*.

of each other: the dying agonies of the surviving mother were sharpened by her reflecting on the forlorn condition in which her children would be left; for, although the Milletts were originally aristocratic and wealthy, the property had undergone so many subdivisions, as to have left but a very slender provision for the member of the family to whom she had united herself.

The affecting appeal which Mrs. Millett is said to have addressed to her sympathising friend, and medical attendant, was not made in vain: on her decease, Mr. Tonkin immediately removed the three children to his own house, and they continued under the guardianship of their kind benefactor, until each, in succession, found a home by marriage.

The eldest sister, Jane, was married to Henry Sampson, a respectable watchmaker at Penzance; the youngest, Elizabeth, to her cousin, Leonard Millett of Marazion; neither of whom had any family. The second sister, Grace, was married to Robert Davy, from which union sprang five children, two boys and three girls, the eldest being Humphry, the subject of our memoir, and the second son, John, now Dr. Davy, a Surgeon to the Forces, and a gentleman distinguished by several papers in the Philosophical Transactions.

Humphry Davy was nursed by his mother, and passed his infancy with his parents;* but his childhood, after they had removed from Penzance to

* For these materials I acknowledge myself indebted to Dr. Penneck of Penzance, and to Mrs. Millett, Sir H. Davy's sister. The facts were communicated in letters to Lady Davy, by whom they were kindly placed at my disposal.

reside on their estate at Varfell, was spent partly with them and partly with Mr. John Tonkin, who extended his disinterested kindness from the mother to all her children, but more especially to Humphry, who is said, when a child, to have exhibited powers of mind superior to his years. I have spared no pains in collecting materials for the illustration of the earlier periods of his history; as, to estimate the magnitude of an object, we must measure the base with accuracy, in order to comprehend the elevation of its summit.

He was first placed at a preparatory seminary kept by a Mr. Bushell, who was so struck with the progress he made, that he urged his father to remove him to a superior school.

It is a fact worthy, perhaps, of being recorded, that he would at the age of about five years turn over the pages of a book as rapidly as if he were merely engaged in counting the number of leaves, or in hunting after pictures; and yet, on being questioned, he could generally give a very satisfactory account of the contents. I have been informed by Lady Davy that the same faculty was retained by him through life, and that she has often been astonished, beyond the power of expression, at the rapidity with which he read a work, and the accuracy with which he remembered it. Mr. Children has also communicated to me an anecdote, which may be related in illustration of the same quality. Shortly after Dr. Murray had published his system of chemistry, Davy accompanied Mr. Children in an excursion to Tonbridge, and the new work was placed in the carriage. During the occasional intervals in which their conversation was suspended, Davy was seen turning over the leaves of the book, but his companion did not believe it possible that he could have made himself acquainted with any part of its contents, until at the close of the journey he surprised him with a critical opinion of its merits.

The book that engaged his earliest attention was "The Pilgrim's Progress," a production well calculated, from the exuberance of its invention, and the rich colouring of its fancy, for seizing upon the ardent imagination of youth. This pleasing work, it will be remembered, was the early and especial favourite of Dr. Franklin, who never alluded to it but with feelings of the most lively delight.

Shortly afterwards, he commenced reading history, particularly that of England; and at the age of eight years he would, as if impressed with the powers of oratory, collect together a number of boys in a circle, and mounting a cart or carriage that might be standing before the inn near Mr. Tonkin's house, harangue them on different subjects, and offer such comments as his own ideas might suggest.

He was, moreover, at this age, a great lover of the marvellous, and amused himself and his schoolfellows by composing stories of romance and tales of chivalry, with all the fluency of an Italian improvisatore; and joyfully would he have issued forth, armed *cap-à-pié*, in search of adventures, and to free the world of dragons and giants.

In this early fondness for fiction, and in the habit of exercising his ingenuity in creating imagery for the gratification of his fancy, Davy and Sir Walter Scott greatly resembled each other. The Author of Waverley, in his general preface to the late edition of his novels, has given us the following account of this talent. "I must refer to a very early period of my life, were I to point out my first achievement as a tale-teller; but I believe most of my old schoolfellows can still bear witness that I had a distinguished character for that talent, at a time when the applause of my companions was my recompense for the disgraces and punishments which the future romance-writer incurred for being idle himself, and keeping others idle, during hours that should have been employed on our tasks." Had not Davy's talents been diverted into other channels, who can say that we might not have received from his inventive pen a series of romantic tales, as beautifully illustrative of the early history of his native country as are the Waverley Novels of that of Scotland? for Cornwall is by no means deficient in elfin sprites and busy "*piskeys*;" the invocation is alone required to summon them from their dark recesses and mystic abodes.

Davy was also in the frequent habit of writing verses and ballads; of making fireworks, and of preparing a particular detonating composition, to which he gave the name of "Thunder-powder," and which he would explode on a stone to the great wonder and delight of his young playfellows.

Another of his favourite amusements may also be recorded in this place; for, however trifling in itself the incident may appear, to the biographer it is full of interest, as tending to show the early existence of that passion for experiment, which afterwards rose so nobly in its aims and objects, as the mind expanded with the advancement of his years. It consisted in scooping out the inside of a turnip, placing a lighted candle in the cavity, and then exhibiting it as a lamp; by the aid of which he would melt fragments of tin, obtained from the metallic blocks which commonly lie about the streets of a coinage town, and demand from his companions a certain number of pins for the privilege of witnessing the operation.

At an early age, but I am unable to ascertain the exact period, he was placed at the Grammar-School in Penzance, under the Rev. J. C. Coryton; and whilst his father resided at Varfell, he lived with Mr. Tonkin, except during the holidays, which he always spent with his parents.

He was extremely fond of fishing; and I have been lately informed by one of his earliest companions, that when very young he greatly excelled in that art. "I have known him," says my correspondent, "catch grey-mullet at Penzance Pier, when none of us could succeed. The mullet is a very difficult fish to hook, on account of the diminutive size of its mouth; but Davy adopted a plan of his own contrivance. Observing that they always swam in shoals, he attached a succession of pilchards to a string, reaching from the surface to the bottom of the sea, and while his prey were swimming around the bait, he would by a sudden movement of the string entangle several of them on the hooks, and thus dexterously capture them."

As soon as he became old enough to carry a gun,

a portion of his leisure hours was passed in the recreation of shooting; a pursuit which also enabled him to form a collection of the rare birds which occasionally frequented the neighbourhood, and which he is said to have stuffed with more than ordinary skill.

When at home, he frequently amused himself with reading and sketching, and sometimes with caricaturing any thing which struck his fancy; on some occasions he would shut himself up in his room, arrange the chairs, and lecture to them by the hour together.

I have been informed by one of his schoolfellows, a gentleman now highly distinguished for his literary attainments, that, in addition to the amusements already noticed, he was very fond of playing at "Tournament," fabricating shields and visors of pasteboard, and lances of wood, to which he gave the appearance of steel by means of black-lead. Thus equipped, the juvenile combatants, like Ascanius and the Trojan youths of classic recollection, would tilt at each other, and perform a variety of warlike evolutions.

By this anecdote we are forcibly reminded of the early taste of Sir William Jones, who, when a boy at Harrow School, invented a political play, in which William Bennet, Bishop of Cloyne, and the celebrated Dr. Parr, were his principal associates. They divided the fields in the neighbourhood of Harrow, according to a map of Greece, into states and kingdoms; each fixed upon one as his dominion, and assumed an ancient name. Some of their schoolfellows consented to be styled Barbarians, who were to invade their territories and attack their hillocks, which they denominated fortresses.*

On one occasion, Davy got up a Pantomime; and I have very unexpectedly obtained a fly-leaf, torn out of a Schrevelius' Lexicon, on which the *Dramatis Personæ,* as well as the names of the young actors, were registered, as originally cast. This document appears so interesting, that I have thought it right to place it on record.

Father	Cunnack.
Harlequin	Davy.
Clown	* * * * * *†
Columbine	Hichens.
Cupid	Veale.
Fortuna	Scobell.
Ben	Billy Giddy.
Nurse	Robyns.
Maccaroni	Dennis.

The performers, who, I believe, with one exception, are all living, will perhaps find some amusement in examining how far their future characters were shadowed forth on this occasion. At all events, I feel confident that they will receive no small gratification at having their recollections thus carried back to the joyous scenes of boyhood, connected as they always are, and must ever be, with the most delightful associations of our lives.

From Penzance school he went to Truro, in the year 1793, and finished his education under the Rev. Dr. Cardew, a gentleman who is distinguished by

* Life of Sir William Jones, by Lord Teignmouth.

† Here, as Mrs. Ratcliffe would say, the Legend is so effaced by damp and time, as to be wholly illegible.

the number of eminent scholars with which he has graced his country.

That he was quick and industrious in his school exercises, may be inferred from an anecdote related by his sister, that "on being removed to Truro, Dr. Cardew found him very deficient in the qualifications for the Class of his age, but on observing the quickness of his talents, and his aptitude for learning, he did not place him in a lower form, telling him that by industry and attention he trusted he might be entitled to keep the place assigned to him; which," his sister says, "he did, to the entire satisfaction of his master."

It is very natural that an anecdote so gratifying to the family should have been deeply imprinted on their memory; but we must not be surprised on finding that it did not make a similar impression upon Dr. Cardew. From a letter lately addressed by that gentleman to Mr. Davies Gilbert, the following is an extract:—"With respect to our illustrious countryman, Sir H. Davy, I fear I can claim but little merit from the share I had in his education. He was not long with me; and while he remained I could not discern the faculties by which he was afterwards so much distinguished; I discovered, indeed, his taste for poetry, which I did not omit to encourage." Dr. Cardew adds, "While engaged in teaching the classics, I was anxious to discharge faithfully the duties of my profession to the best of my ability; but I was certainly fortunate in having so many good materials to work upon, and thus having only '*fungi vice cotis,*' though '*exsors ipse secandi.*'"—To the truth of this latter part of the Doctor's quotation, will his scholars willingly subscribe? It may be fairly doubted how far Dr. Cardew was able to descend into the shadowy regions of Maro, without the "*donum fatalis virgæ.*"

Mrs. Millett thinks that the deficiency just alluded to may be attributed to Mr. Coryton, rather than to the inattention of her brother; the former having, from his neglect as a master, given very general dissatisfaction. From what I can learn, at this distant period, of the character of Mr. Coryton, it appears at all events, that the "*exsors ipse secandi*" could not have been justly applied to him; and that, owing to an unfortunate aptness in the name to a doggrel verse, poor Davy had frequently to smart under his tyranny.

"Now, Master Dàvy,
Now, Sir, I hàve 'e,
No one shall sàve 'e,
Good Master Dàvy;"

when the master, suiting the action to the rhythm, inflicted upon the hand of the unlucky scholar the verberations of that type and instrument of pedagoguish authority—the flat ruler. Here we have another example of the seduction of sound, argued by our great jurist Mr. Bentham,* to have determined the maxims of that law, which has been pronounced by its sages the perfection of reason.

* "Were the enquiry diligently made," he says, "it would be found that the Goddess of Harmony has exercised more influence, however latent, over the dispensations of Themis, than her most diligent historiographers, or even her most passionate panegyrists, seem to be aware of. Every one knows how, by the ministry of

From a letter, however, written by Davy a few years afterwards, respecting the education of a member of his family, he would appear to have entertained an opinion not very unlike that of John Locke; for, although he testifies the highest respect for Dr. Cardew, he seems to consider the comparative idleness of his earlier school career, by allowing him to follow the bent of his own mind, to have favoured the developement of his peculiar genius. "After all," he says, "the way in which we are taught Latin and Greek, does not much influence the important structure of our minds. I consider it fortunate that I was left much to myself as a child, and put upon no particular plan of study, and that I enjoyed much idleness at Mr. Coryton's school. I perhaps owe to these circumstances the little talents I have, and their peculiar application:—what I am I have made myself—I say this without vanity, and in pure simplicity of heart."

His temper during youth is represented as mild and amiable. He never suppressed his feelings, but every action was marked by ingenuousness and candour, qualities which endeared him to his youthful associates, and gained him the love of all who knew him. "Nor can I find," says his sister, "beloved as he must have been by my mother, that she showed him any particular preference; — all her children appeared to be alike her care, and all alike shared her affection."

Orpheus, it was she who first collected the sons of men beneath the shadow of the sceptre: yet in the midst of continual experience, men seem yet to learn with what successful diligence she laboured to guide it in its course."

In 1794, Mr. Davy died. We cannot but regret that he did not live long enough to witness his son's eminence; for life, as Johnson says, has few better things to give than a talented son; but from his widow, who has but lately descended to the tomb, full of years and respectability, this boon was not withheld, she witnessed his whole career of usefulness and honour, and happily closed her eyes before her maternal fears could have been awakened by those signs of premature decay, which for some time had excited in his friends, and in the friends of science, an alarm which the recent deplorable event has too fatally justified.

In the year following the decease of her husband, Mrs. Davy, who had again taken up her residence in Penzance, apprenticed her son,* by the advice of her long-valued friend, Mr. Tonkin, to Mr. John Bingham Borlase, at that time a surgeon and apothecary, but who afterwards obtained a diploma, and became an eminent physician at Penzance. Davy, however, for the most part, continued to pursue his own plans of study; for although his friend Mr. Tonkin, without doubt, intended him for a general practitioner in his native town, yet he himself always looked forward to graduation at Edinburgh, as a preliminary measure to his practising in the higher walk of his profession.

His mind had, for some time, been engrossed with philosophical pursuits; but until after he had been placed with Mr. Borlase, it does not appear that he indicated any decided turn for chemistry, the study of

* The original indenture, now in the possession of Mr. R. Edmonds, solicitor, of Penzance, is dated February 10th, 1795.

which he then commenced with all the ardour of his temperament; and his eldest sister, who acted as his assistant, well remembers the ravages committed on her dress by corrosive substances.

It has been said that his mind was first directed to chemistry by a desire to discover various mixtures as pigments: a suggestion to which, I confess, I am not disposed to pay much attention; for although he might have sought by new combinations to impart a novel and vivid richness of colouring to his drawings, it was the character of his mind to pursue with ardour every subject of novelty, and to get at results by his own native powers, rather than by the recorded experience of others.

I must here relate an anecdote, in illustration of this statement, which has been lately communicated to me by the Reverend Dr. Batten, the principal of the East India College at Hayleybury. This gentleman was one of the earliest of Davy's schoolfellows, but as he advanced in age, different views, and a different plan of education, carried him to a distant part of the kingdom; the discipline and duties of a cloistered school necessarily estranged him from his native town; and it was not until after his admission at Cambridge, and the arrival of the long vacation, which afforded a temporary oblivion of academic cares, that Mr. Batten returned to Cornwall, to revisit the scenes, and to renew the friendships of his boyish days. Davy, who was at that period an apprentice to Mr. Borlase, received him with transport and affection; but he was no longer the boy that his friend had left him; he had become more serious and contemplative, fond of solitary rambles, and averse to enter into society, or to join the festive parties of the inhabitants. In fact, his mind was now in the act of being moulded by the spirit of Nature; and, without the constraint of study, he was insensibly inhaling knowledge with the wild breezes of his native hills.

In the course of conversation, Mr. Batten spoke of his academic studies; and in alluding to the principles of Mechanics, to which he had lately paid much attention, he expressed himself more particularly pleased with that part which treats of "the Collision of Bodies." What was his surprise, on finding Davy as well, if not better acquainted with its several propositions! It was true that he had never systematically studied the subject—had never perhaps seen any standard work upon it, but he had instituted experiments with elastic and inelastic balls, and had worked out the results by the unassisted energies of his own mind. It is clear that, had this branch of science not existed, Davy would have created it.

During this period of his apprenticeship, he twice a week attended a French school in Penzance, kept by a M. Dugast, a priest from La Vendée; and it was remarked that, although he acquired a knowledge of the grammatical construction of the language with greater facility than any of the other scholars, he could not succeed in obtaining the pronunciation; and, in fact, notwithstanding his extensive intercourse with foreigners, and his residence in France, he never, even in after life, could pronounce French with correctness or speak it with fluency.

While with Mr. Borlase, it was his constant custom to walk in the evening to Marazion, to drink tea with an aunt to whom he was greatly attached. Upon such occasions, his usual companion was a hammer, with which he procured specimens from the rocks on the beach. In short, it would appear that, at this period, he paid much more attention to Philosophy than to Physic; that he thought more of the bowels of the earth, than of the stomachs of his patients; and that, when he should have been bleeding the sick, he was opening veins in the granite. Instead of preparing medicines in the surgery, he was experimenting in Mr. Tonkin's garret, which had now become the scene of his chemical operations; and, upon more than one occasion, it is said that he produced an explosion, which put the Doctor, and all his glass bottles, in jeopardy. "This boy Humphry is incorrigible!"—"Was there ever so idle a dog!"—"He will blow us all into the air!" Such were the constant exclamations of Mr. Tonkin; and then, in a jocose strain, he would speak of him as the "Philosopher," and sometimes call him "Sir Humphry," as if prophetic of his future renown.*

His sister has remarked that, as he advanced in life, he always preferred the society of persons older than himself; and one of his contemporaries informs me that he never heard him allude to any subject of science, although he remembers that while one of his pockets was filled with fishing-tackle, the

* Davy appears to have been more fortunate than his prototype Scheele; for on one occasion, as the latter was employed in making pyrophorus, a fellow apprentice, without his knowledge, put some fulminating powder into the mixture; the consequence was a violent explosion; the whole family was thrown into confusion, and the young chemist was severely chastised.

other was as commonly loaded with specimens of rocks. With those, however, who were superior to him in years, he delighted to enter into discussion. At Penzance, there still resides a member of the Society of Friends, whose ingenuity entitles him to greater rewards than a provincial town can afford, with whom Davy, as a boy, was in the constant habit of discussing questions of practical mechanics. "I tell thee what, Humphry," exclaimed the Quaker upon one of these occasions—"thou art the most quibbling hand at a dispute I ever met with in my life."

For the surgical department of the profession, he always entertained a decided distaste, although the following extract from a letter of my correspondent Mr. Le Grice will show that, for once at least, he had the merit of mending a broken head. "The first time I ever saw Davy was on the Battery rocks; we were alone bathing, and he pointed out to me a good place for diving; at the same time he talked about the tides, and Sir Isaac Newton, in a manner that greatly amazed me. I perhaps should not have so distinctly remembered him, but on the following day, by not exactly marking the spot he had pointed out, I was nearly killed by diving on a rock, and he came as Mr. Borlase's assistant to dress the wound."

It was his great delight to ramble along the seashore, and often, like the orator of Athens, would he on such occasions declaim against the howling of the wind and waves, with a view to overcome a defect in his voice, which, although only slightly perceptible in his maturer age, was in the days of his boyhood exceedingly discordant. I may be allowed to observe, that the peculiar intonation he employed in his public addresses, and which rendered him obnoxious to the charge of affectation, was to be referred to a laborious effort to conceal this natural infirmity. It was also clear that he was deficient in that quality which is commonly called "a good ear," and with which the modulation of the voice is generally acknowledged to have an obvious connexion. Those who knew him intimately will readily bear testimony to this fact. Whenever he was deeply absorbed in a chemical research, it was his habit to hum some tune, if such it could be called, for it was impossible for any one to discover the air he intended to sing: indeed, Davy's music became a subject of raillery amongst his friends; and Mr. Children informs me, that, during an excursion, they attempted to teach him the air of 'God save the King,' but their efforts were unavailing.

It may be a question how far the following fact, with which I have just been made acquainted, admits of explanation upon this principle. On entering a volunteer infantry corps, commanded by a Captain Oxnam, Davy could never emerge from the awkward squad; no pains could make him keep the step; and those who were so unfortunate as to stand before him in the ranks, ought to have been heroes invulnerable in the heel. This incapacity, as may be readily supposed, occasioned him considerable annoyance, and he engaged a serjeant to give him private lessons, but it was all to no purpose. In the platoon exercise he was not more expert; and he whose electric battery was destined to triumph over the animosity of nations, could never be taught to shoulder a musket in his native town.

That Davy, in his youth, possessed courage and decision, may be inferred from the circumstance of his having, upon receiving a bite from a dog supposed to be rabid, taken his pocket-knife, and without the least hesitation cut out the part on the spot, and then retired into the surgery and cauterized the wound; an operation which confined him to Mr. Tonkin's house for three weeks. The gentleman from whom I received an account of this adventure, the accuracy of which has been since confirmed by Davy's sister, also told me, that he had frequently heard him declare his disbelief in the existence of pain whenever the energies of the mind were directed to counteract it; but he added, "I very shortly afterwards had an opportunity of witnessing a practical refutation of this doctrine in his own person; for upon being bitten by a conger eel, my young friend Humphry roared out most lustily."

The anecdote of Davy's excising the bitten part with so much promptitude and coolness, derives its interest from the age and inexperience of the operator. In the course of his practice, every physician must have met with similar cases of stern decision; but I will venture to say that they have never occurred except in instances of persons of acknowledged courage. Not many days since, a veteran officer, distinguished for the intrepidity with which he rescued the person of George the Third from the fury of a desperate mob, in St. James's Park, informed me that he had formerly been bitten at Vienna by a dog afterwards ascertained to have been rabid; he immediately entered a blacksmith's shop, and by threats compelled the person at the forge to heat an iron red-hot, and burn his leg to the bone. The blacksmith, after first stipulating that he should strap his eccentric customer to the anvil, reluctantly complied; and my friend showed me a scar which sufficiently testified the complete manner in which the son of Vulcan had performed his engagement:—But to return from this digression.

At this time of day, no one can surely believe with Pope, that a "Ruling Passion" is an innate and irresistible affection antecedent to reason and observation: on the contrary, ample experience has led us to the conclusion, that

> ————"men's judgments are
> A parcel of their fortunes, and things outward
> Do draw the inward quality after them."

The prevailing bias of great minds may thus be often traced to some accidental, and apparently trivial, impression in early life; and the acute biographer, in the course of his observations, will continually discover traits of character that are readily referable to such a source, even as in the magical colouring of Rembrandt's works, the practised eye will recognize the *chiaro-oscuro* of his father's mill, in which the artist passed his hours of childhood.

In like manner, that marked aversion to arbitrary power, which ever distinguished the actions and writings of Dr. Franklin, has by himself been referred to the sense of injustice early imprinted upon his mind by the severe and tyrannical conduct of his elder brother; while, at the same time, he tells us that he was indebted for his habit through life, of forming just estimates of the value of things, to his having, at the age of seven years, "paid too much for his whistle."

But circumstances, however disposed and happily combined, although they may direct, can never create genius; it is possible that Cowley might never have been enamoured of the Muses, nor Sir Joshua Reynolds have courted the Graces, but for the casual circumstances recorded by the biographer; and Ferguson might not have turned his attention to mechanical inventions, had not an accident befallen the roof of his father's cottage; and even Priestley, the founder of a new and beautiful department in science, might very probably never have been led to think of pneumatic chemistry, had he not lived in the vicinity of a great brewery: still, however, such men could not have shone dimly, if true genius be correctly defined by Dr. Johnson as "a mind of large general powers accidentally determined to some particular direction."*—So with Davy; his mind was as vigorous as it was original, and no less logical and precise than it was daring and comprehensive; nothing was too mighty for its grasp, nothing too minute for its observation; like the trunk of the

* M. de Bourrienne, in his "Private Memoirs of Napoleon Buonaparte," appears to have justly appreciated the influence of circumstances upon the destinies of great men. In speaking of Buonaparte at the Military College of Brienne, he says, "If the monks, to whom the superintendence of the establishment was confided, had engaged more able mathematical professors, or if we had had any excitement to the study of Chemistry, or Natural Philosophy, I am convinced that Buonaparte would have pursued those sciences with all the genius and spirit of investigation, which he displayed in a career more brilliant, it is true, but far less useful to mankind."

elephant, it could tear up the oak of the forest, or gently pluck the acorn from its branch.

That circumstances in early life should have directed such energies to a science, which requires for its advancement all the aids of novel and bold, and yet patient and accurate research, is one of those fortunate events which every unprejudiced mind will view with triumph.

It is surely not difficult to understand how it happened that a mind endowed with the genius and sensibilities of Davy, should have been directed to the study of Chemistry and Mineralogy, when we consider the nature and scenery of the country in which accident had placed him. Many of his friends and associates must have been connected with mining speculations: "Shafts," "Cross Courses," and "Lodes," were words familiarised to his ears; and his native love of enquiry could not have long suffered them to remain strangers to his understanding. Nor could he have wandered along the rocky coast, or have reposed for a moment to contemplate its wild scenery, without being invited to geological enquiry by the genius of the place; for were we to personify the science, where could we find a more appropriate spot for her "local habitation?" "How often when a boy," said Davy to me, on my showing him a drawing of the wild rock scenery of Botallack Mine, "have I wandered about those rocks in search of new minerals, and, when fatigued, sat down upon the turf, and exercised my fancy in anticipations of scientific renown!"

Such scenery, also, in one who possessed a quick sensibility to the sublime forms of Nature, was well calculated to kindle that enthusiasm which is so essential to poetical genius; and we accordingly learn, that he became enamoured of the Muses at a very early age, and evinced his passion by several poetical productions. I am assured by Dr. Batten that, at the age of twelve years, he had finished an epic poem, which he entitled the "Tydidiad," from its celebrating the adventures of Diomede on his return from the Trojan war. It is much to be regretted that not even a fragment of this poem should have been preserved; but Dr. Batten well remembers that it was characterised by great freedom of invention, vigour of description, and wildness of execution.

At the age of seventeen he became desperately enamoured of a young French lady, at that time resident at Penzance, to whom he addressed numerous sonnets; but these, like the passion that produced them, have long since been extinct.

Several of his minor productions were printed in a work entitled the "Annual Anthology," published in three volumes at Bristol, in 1799; two of which were edited by Southey, and one by James Tobin;—a work of some curiosity, independent of its merits, as the first attempt in this country to establish an "Annual," a species of literary composition which has lately been made very popular and amusing.

These volumes have now become extremely scarce, for which, and other reasons, I have thought it right to place Davy's productions on record in these memoirs; for although they are marked by the common faults of youthful poets, they still bear the stamp of lofty genius. There is, besides, a vein of philosophical contemplation running through their composition, which may be considered as indicating the future character and pursuit of their author; an ardent aspiration after fame seems, even at this early period, to have been felt in all its force, and is expressed in many striking and beautiful passages.

There is still a higher motive by which I am induced to introduce these specimens into my memoir, that of showing the bias of his genius at this early period, with a view to compare it with that which displayed itself in the "last days of the philosopher." We shall find that the bright and rosy hues of fancy which gilded the morning of his life, and were subdued or chased away by the more resplendent light of maturer age, again glowed forth in the evening of his days, and illumined the setting, as they had the dawning of his genius.

His first production bears the date of 1795, and is entitled

THE SONS OF GENIUS.

Bright bursting through the awful veil of night
  The lunar beams upon the Ocean play,
The watery billows shine with trembling light,
  Where the swift breezes skim along the sea.

The glimmering stars in yon ethereal plain
  Grow pale, and fade before the lucid beams,
Save where fair Venus, shining o'er the main
  Conspicuous, still with fainter radiance gleams.

Clear is the azure firmament above,
  Save where the white cloud floats upon the breeze,
All tranquil is the bosom of the grove,
  Save where the Zephyr warbles through the trees.

Now the poor shepherd wandering to his home
Surveys the darkening scene with fearful eye,
On every green sees little elfins roam,
And haggard sprites along the moonbeams fly.

While Superstition rules the vulgar soul,
Forbids the energies of man to rise,
Raised far above her low, her mean controul,
Aspiring Genius seeks her native skies.

She loves the silent solitary hours,
She loves the stillness of the starry night,
When o'er the brightening view Selene pours
The soft effulgence of her pensive light.

Tis then disturb'd not by the glare of day;
To mild tranquillity alone resign'd,
Reason extends her animating sway
O'er the calm empire of the peaceful mind.

Before her lucid, all-enlightening ray,
The pallid spectres of the Night retire,
She drives the gloomy terrors far away,
And fills the bosom with celestial fire.

Inspired by her, the Sons of Genius rise
Above all earthly thoughts, all vulgar care;
Wealth, power, and grandeur they alike despise,
Enraptured by the good, the great, the fair.

A thousand varying joys to them belong—
The charms of Nature and her changeful scenes;
Their's is the music of the vernal song,
And their's the colours of the vernal plains.

Their's is the purple-tinged evening ray,
With all the radiance of the morning sky;
Their's is the splendour of the risen day,
Enshrined in glory by the sun's bright eye.

For them the Zephyr fans the odorous gale,
  For them the warbling streamlet softly flows,
For them the Dryads shade the verdant vale,
  To them sweet Philomel attunes her woes.

To them no wakeful moonbeam shines in vain
  On the dark bosom of the trackless wood,
Sheds its mild radiance o'er the desert plain,
  Or softly glides along the chrystal flood.

Yet not alone delight the soft and fair,
  Alike the grander scenes of Nature move;
Yet not alone her beauties claim their care,
  The great, sublime, and terrible, they love.

The Sons of Nature, they alike delight
  In the rough precipice's broken steep,
In the black terrors of the stormy night,
  And in the thunders of the threatening deep.

When the red lightnings through the ether fly,
  And the white foaming billows lash the shores;
When to the rattling thunders of the sky
  The angry Demon of the waters roars;

And when, untouch'd by Nature's living fires,
  No native rapture fills the drowsy soul;
Then former ages, with their tuneful lyres,
  Can bid the fury of the passions fall.

By the blue taper's melancholy light,
  Whilst all around the midnight torrents pour,
And awful glooms beset the face of Night,
  They wear the silent solitary hour.

Ah, then, how sweet to pass the night away
  In silent converse with the Grecian page!
Whilst Homer tunes his ever-living lay,
  Or reason listens to th' Athenian sage;

To scan the laws of Nature, to explore
  The tranquil reign of mild Philosophy ;
Or on Newtonian wings sublime to soar
  Through the bright regions of the starry sky.

Ah ! who can paint what raptures fill the soul
  When Attic Freedom rises to the war,
Bids the loud thunders of the battle roll,
  And drives the tyrant trembling from her shore !

From these pursuits the Sons of Genius scan
  The end of their creation ; hence they know
The fair, sublime, immortal hopes of man,
  From whence alone undying pleasures glow.

By Science calm'd, over the peaceful soul,
  Bright with eternal Wisdom's lucid ray,
Peace, meek of eye, extends her soft controul,
  And drives the fury Passions far away.

Virtue, the daughter of the skies supreme,
  Directs their life, informs their glowing lays—
A steady friend ; her animating beam
  Sheds its soft lustre o'er their latter days.

When life's warm fountains feel the frost of time ;
  When the cold dews of darkness close their eyes,
She shows the parting soul, upraised sublime,
  The brighter glories of her kindred skies.

Thus the pale Moon, whose pure celestial light
  Has chased the gloomy clouds of Heaven away,
Rests her white cheek, with silver radiance bright,
  On the soft bosom of the Western sea.

Lost in the glowing wave, her radiance dies ;
  Yet, while she sinks, she points her ling'ring ray
To the bright azure of the orient skies—
  To the fair dawning of the glorious day.

Like the tumultuous billows of the sea
  Succeed the generations of mankind;
Some in oblivious silence pass away,
  And leave no vestige of their lives behind.

Others, like those proud waves which beat the shore,
  A loud and momentary murmur raise;
But soon their transient glories are no more,—
  No future ages echo with their praise.

Like yon proud rocks amidst the sea of time,
  Superior, scorning all the billows' rage,
The living Sons of Genius stand sublime,
  Th' immortal children of another age.

For those exist whose pure ethereal minds,
  Imbibing portions of celestial day,
Scorn all terrestrial cares, all mean designs,
  As bright-eyed eagles scorn the lunar ray.

Their's is the glory of a lasting name,
  The meed of Genius and her living fires,
Their's is the laurel of eternal fame,
  And their's the sweetness of the Muse's lyres.

D.—1795

THE SONG OF PLEASURE.

The genial influence of the day
Had chased the lingering cold away;
Borne upon the Zephyr's wing,
Sweetly smiled the radiant Spring:
Her mild re-animating breath
Wakes Nature from her wintry death;
Attended by the laughing Hours,
She rises clad in flowers,
And lightly as she trips along,
The vernal warblers raise the song.

Rich in a thousand radiant dyes,
Around her steps the flow'rets rise,
The Zephyr sports, the sunbeams sleep
On the blue bosom of the deep.
And now, within my throbbing breast
I feel the influence of the Spring,
To ecstasy I tune my string,
And garlanded with odorous flowers,
I hasted to the shady grove,
I hasted to the roseate bowers
Where Pleasure dwells with Love.

There Youth, and Love, and Beauty, bound
The glowing rose my harp around;
Then to the daughter of Desire,
To bright-eyed Pleasure gave the lyre:
She tuned the string,
And smiling softer than the rosy sea,
When the young Morning blushes on her breast,
She raised the raptured lay,
I heard her sing,
The song lull'd every care and every thought to rest.

Sons of Nature, hither haste,
The blessings of existence taste;
Listen to my friendly lay,
And your cares shall fly away,
Quick as fly the wintry snows
When the vernal Zephyr blows.
Let others, courting war's alarms,
Seek the bloody field of arms;
Let others, with undaunted soul,
Bid Bellona's thunders roll;
From the lightnings of their eye
Let the trembling squadrons fly;
Sons of Nature, you shall prove
A softer fight, the fight of love.

While you in soft repose are laid
Underneath the myrtle shade,
Amid the murky glooms of Death,
The sons of battle pant for breath.

Let the philosophic sage,
His silver tresses white with age,
Amid the chilling midnight damp,
Waste the solitary lamp,
To scan the laws of Nature o'er,
The paths of Science to explore;
Curb'd beneath his harsh controul
The blissful Passions fly the soul.
You, the gentler sons of joy,
Softer studies shall employ!
He to curb the Passions tries,
You shall bid them all arise;
His wants he wishes to destroy,
You shall all your wants enjoy.
Let the laurel, Virtue's meed,
Crown his age-besilver'd head,
The verdant laurel ever grows
Amid the sullen Winter's snows:
Let the rose, the flower of bliss,
The soft unwrinkled temples kiss;
Fann'd by the Zephyr's balmy wing,
The odorous rose adorns the Spring.

Let the Patriot die, to raise
A lasting monument of praise.
Ah, fool, to tear the glowing rose
From the mirth-encircled brows,
That around his dusky tomb
The ever-verdant bay may bloom!
Let Ambition's sons alone
Bow around the tottering throne,
Fly at Glory's splendid rays,
And, moth-like, die amidst a blaze;

You shall bow, and bow alone,
Before delicious Beauty's throne.
Lo! Theora treads the green,
All breathing grace and harmony she moves,
Fair as the mother of the Loves.
In graceful ringlets floats her golden hair;
From the bright azure of her eye
Expression's liquid lightnings fly.
Her cheek is fair,
Fair as the lily, when at dawning day,
Tinged with the morning's bright and purple ray,
Yonder scented groves among
She will listen to your song.

In yonder bower where roses bloom,
Where the myrtle breathes perfume,
You shall at your ease recline,
And sip the soul-enlivening wine;
There the lyre, with melting lay,
Shall bid the soul dissolve away.
Soft as the Morning sheds her purple light
Through the dark azure of the Night,
So soft the God of slumber sheds
His roseate dews around your heads.

Such the blessings I bestow!
Haste, my sons, these blessings know!
Behold the flow'rets of the Spring,
They wanton in the Zephyr's wing,
They drink the matin ether blue,
They sip the fragrant evening dew.
Man is but a short-lived flower,
His bloom but for a changeful hour!
Pass a little time away,
The rosy cheek is turn'd to clay:
No living joys, no transports burn
In the dark sepulchral urn,
No *Laurels* crown the fleshless brows,
They fade together with the *Rose*. D.—1796.

ODE TO SAINT MICHAEL'S MOUNT, IN CORNWALL.

THE sober eve with purple bright
Sheds o'er the hills her tranquil light
  In many a lingering ray;
The radiance trembles on the deep,
Where rises rough thy rugged steep,
  Old Michael, from the sea.

Around thy base, in azure pride,
Flows the silver-crested tide,
  In gently winding waves;
The Zephyr creeps thy cliffs around,—
Thy cliffs, with whispering ivy crown'd,
  And murmurs in thy caves.

Majestic steep! Ah, yet I love,
With many a lingering step, to rove
  Thy ivied rocks among;
Thy ivied, wave-beat rocks recall
The former pleasures of my soul,
  When life was gay and young

Enthusiasm, Nature's child,
Here sung to me her wood-songs wild,
  All warm with native fire;
I felt her soul-awakening flame,
It bade my bosom burn for fame,—
  It bade me strike the lyre.

Soft as the Morning sheds her light
Through the dark azure of the Night
  Along the tranquil sea;
So soft the bright-eyed Fancy shed
Her rapturing dreams around my head,
  And drove my cares away.

When the white Moon with glory crown'd,
The azure of the sky around,
  Her silver radiance shed;
When shone the waves with trembling light,
And slept the lustre palely bright
  Upon thy tower-clad head;

Then BEAUTY bade my pleasure flow,—
Then BEAUTY bade my bosom glow,
  With mild and gentle fire!
Then Mirth, and Cheerfulness, and Love,
Around my soul were wont to move,
  And thrill'd upon my lyre.

But when the Demon of the deep
Howl'd around thy rocky steep,
  And bade the tempests rise,—
Bade the white foaming billows roar,
And murmuring dash the rocky shore,
  And mingle with the skies;

Ah, then my soul was raised on high,
And felt the glow of ecstasy,
  With *great* emotions fill'd;
Thus Joy and Terror reign'd by turns,
And now with LOVE the bosom burns,
  And now by FEAR is chill'd.

Thus to the sweetest dreams resign'd,
The fairy FANCY ruled my mind,
  And shone upon my youth;
But now, to awful Reason given,
I leave her dear ideal heaven
  To hear the voice of TRUTH.

She claims my best, my loftiest song,
She leads a brighter maid along—

DIVINE PHILOSOPHY,
Who bids the mounting soul assume
Immortal Wisdom's eagle plume,
And penetrating eye,

Above Delusion's dusky maze,
Above deceitful Fancy's ways,
With roses clad to rise;
To view a gleam of purest light
Bursting through Nature's misty night,—
The radiance of the skies.

D.—1796.

THE TEMPEST.

THE Tempest has darken'd the face of the skies,
The winds whistle wildly across the waste plain,
The Fiends of the whirlwind terrific arise,
And mingle the clouds with the white-foaming main.

All dark is the night, and all gloomy the shore,
Save when the red lightnings the ether divide,
Then follows the thunder with loud-sounding roar,
And echoes in concert the billowy tide.

But though now all is murky and shaded with gloom,
Hope, the soother, soft whispers the tempests shall cease;
Then Nature again in her beauty shall bloom,
And enamour'd embrace the fair sweet-smiling Peace;

For the bright-blushing morning, all rosy with light,
Shall convey on her wings the Creator of day;
He shall drive all the tempests and terrors of night,
And Nature enliven'd, again shall be gay.

Then the warblers of Spring shall attune the soft lay,
And again the bright flow'ret shall blush in the vale;
On the breast of the Ocean the Zephyr shall play,
And the sunbeam shall sleep on the hill and the dale.

If the tempests of Nature so soon sink to rest—
If her once-faded beauties so soon glow again,
Shall Man be for ever by tempests oppress'd,
By the tempests of passion, of sorrow, and pain?

Ah, no! for his passions and sorrow shall cease
When the troublesome fever of life shall be o'er;
In the night of the grave he shall slumber in peace,
And passion and sorrow shall vex him no more.

And shall not this night and its long dismal gloom,
Like the night of the tempest, again pass away?
Yes! the dust of the earth in bright beauty shall bloom,
And rise to the morning of heavenly day!

D.—1796.

EXTRACT FROM AN UNFINISHED POEM ON MOUNT'S BAY.

Mild blows the Zephyr o'er the Ocean dark,
The Zephyr wafting the grey twilight clouds
Across the waves, to drink the solar rays
And blush with purple.
By the orient gleam
Whitening the foam of the blue wave that breaks
Around his granite feet, but dimly seen,
Majestic Michael rises. He whose brow
Is crown'd with castles, and whose rocky sides
Are clad with dusky ivy: he whose base,
Beat by the storm of ages, stands unmoved
Amidst the wreck of things, the change of time.
That base encircled by the azure waves,
Was once with verdure clad: the tow'ring oaks
There waved their branches green,—the sacred oaks
Whose awful shades among, the Druids stray'd
To cut the hallow'd miseltoe, and hold
High converse with their Gods.

On yon rough crag,
Where the wild Tamarisk whistles to the sea blast,
The Druid's harp was heard, swept by the breeze
To softest music, or to grander tones
Awaken'd by the awful master's hand.
Those tones shall sound no more! the rushing waves,
Raised from the vast Atlantic, have o'erwhelm'd
The sacred groves. And deep the Druids lie
In the dark mist-clad sea of former time.
Ages had pass'd away, the stony altar
Was white with moss, when on its rugged base
Dire Superstition raised the gothic fane,
And monks and priests existed.

On the sea
The sunbeams tremble; and the purple light
Illumes the dark Bolerium,* seat of storms.
High are his granite rocks. His frowning brow
Hangs o'er the smiling Ocean. In his caves
Th' Atlantic breezes murmur. In his caves,
Where sleep the haggard Spirits of the storm,
Wild dreary are the *schistine* † rocks around
Encircled by the wave, where to the breeze
The haggard Cormorant shrieks. And far beyond
Are seen the cloud-like Islands, grey in mists.‡

Thy awful height, Bolerium, is not loved
By busy Man, and no one wanders there
Save he who follows Nature,—he who seeks
Amidst thy crags and storm-beat rocks to find
The marks of changes teaching the great laws
That raised the globe from chaos; or he whose soul

---

* The Land's End in Cornwall.

† The granite of Cornwall is generally found incumbent on primitive *schistus*. This is the case in many of the cliffs at the Land's End. The upper stratum is composed of granite, the lower with the surrounding rocks of *schistus*. D.

‡ The Islands of Scilly.

Is warm with fire poetic,—he who feels
When Nature smiles in beauty, or sublime
Rises in majesty,—he who can stand
Unawed upon thy summit, clad in tempests,
And view with raptured mind the roaring deep
Rise o'er thy foam-clad base, while the black cloud
Bursts with the fire of Heaven—
He whose heart
Is warm with love and mercy,—he whose eye
Drops the bright tear when anxious Fancy paints
Upon his mind the image of the Maid,
The blue-eyed Maid who died beneath thy surge.
Where yon dark cliff* o'ershadows the blue main,
THEORA died amidst the stormy waves,
And on its feet the sea-dews wash'd her corpse,
And the wild breath of storms shook her black locks.
Young was THEORA; bluer was her eye
Than the bright azure of the moonlight night;
Fair was her cheek as is the ocean cloud
Red with the morning ray.
Amidst the groves,
And greens, and nodding rocks that overhang
The grey Killarney, pass'd her morning days
Bright with the beams of joy.
To solitude,
To Nature, and to God, she gave her youth;
Hence were her passions tuned to harmony.
Her azure eye oft glisten'd with the tear
Of sensibility, and her soft cheek
Glow'd with the blush of rapture. Hence, she loved
To wander 'midst the green-wood, silver'd o'er
By the bright moonbeam. Hence, she loved the rocks
Crown'd with the nodding ivy, and the lake
Fair with the purple morning, and the sea
Expansive mingling with the arched sky.

---

* A rock near the Land's End, called the 'Irish Lady.'

Kindled by Genius, in her bosom glow'd
The sacred fire of Freedom. Hence, she scorn'd
The narrow laws of custom that control
Her feeble sex. Great in her energies,
She roam'd the fields of Nature, scann'd the laws
That move the ruling atoms, changing still,
Still rising into life. Her eagle eye,
Piercing the blue immensity of space,
Held converse with the lucid sons of Heaven,
The day-stars of creation, or pursued
The dusky planets rolling round the Sun,
And drinking in his radiance light and life.
Such was the Maiden! Such was she who fled
Her native shores.

Dark in the midnight cloud,
When the wild blast upon its pinions bore
The dying shrieks of Erin's injured sons,*
She 'scaped the murderer's arm.

The British bark
Bore her across the ocean. From the West
The whirlwind rose, the fire-fraught clouds of Heaven
Were mingled with the wave. The shatter'd bark
Sunk at thy feet, Bolerium, and the white surge
Closed on green Erin's daughter.

That the Genius who presided over the destinies of Davy should have torn him from these flowery regions of Fancy, and condemned him to labour in the dusky caverns of the mineral kingdom, has furnished a fruitful theme of lamentation to the band of Poets, and to those who prefer the amusements to the profits of life, and who cherish the hallucinations

* The Irish Lady was shipwrecked at the Land's End, about the time of the massacre of the Irish Protestants by the Catholics, in the reign of Charles the First.

of the imagination rather than the truths of science. If, however, we regret that Davy's Muse, like Proserpine, should have been thus violently seized, and carried off to the lower regions, as she was weaving her native wild flowers into a garland, we may console ourselves in knowing that, like the daughter of Ceres, she also obtained the privilege of occasionally revisiting her native bowers; for it will appear in the course of these memoirs, that in the intervals of more abstruse studies, Davy not unfrequently amused himself with poetical composition. But, in sober truth, is it possible that any reasonable being can regret the course in which he has been impelled? A great poetic Genius has said, "If Davy had not been the first Chemist, he would have been the first Poet of his age." Upon this question I do not feel myself a competent judge: but where is the modern Esau who would exchange his Bakerian Lecture for a poem, though it should equal in design and execution the PARADISE LOST?

As far as can be ascertained, one of the first original experiments in Chemistry performed by him at Penzance, was for the purpose of discovering the quality of the air contained in the bladders of sea-weed, in order to obtain results in support of a favourite theory of light; and to ascertain whether, as land vegetables are the renovators of the atmosphere of land-animals, sea-vegetables might not be the preservers of the equilibrium of the atmosphere of the ocean. From these experiments he concluded, that the different orders of the marine *Cryptogamia* were capable of decomposing water, when assisted by the attraction of light for oxygen.

His instruments, however, were of the rudest description, manufactured by himself out of the motley materials which chance threw in his way; the pots and pans of the kitchen, and even the more sacred vessels and professional instruments of the surgery, were without the least hesitation or remorse put in requisition.

While upon this subject, I will relate an anecdote which was communicated to me by my late venerable friend Mr. Thomas Giddy.* A French vessel having been wrecked off the Land's End, the surgeon escaped, and found his way to Penzance; accident brought him acquainted with Humphry Davy, who showed him many civilities, and in return received, as a present from the surgeon, a case of instruments which had been saved from the ship. The contents were eagerly turned out and examined by the young chemist, not, however, with any professional view as to their utility, but in order to ascertain how far they might be convertible to experimental purposes. The old-fashioned and clumsy glyster apparatus was viewed with exultation, and seized in triumph!—What reverses may not be suddenly effected by a simple accident! so says the moralist. Reader, behold an illustration:—in the brief space of an hour, did this long-neglected and unobtrusive machine, emerging from its obscurity and insignificance, figure away in all the pomp and glory of a complicated piece of pneumatic apparatus: nor did its fortunes

* I cannot allude to this name, without paying a tribute of respect to the memory of one who, for more than half a century, practised the profession of a surgeon in Penzance with as much credit to himself, as advantage to his neighbourhood.

end here; it was destined for greater things; and we shall hereafter learn that it actually performed the duties of an air-pump, in an original experiment on the nature and sources of heat. The most humble means may certainly accomplish the highest ends: the filament of a spider's web has been used to measure the motions of the stars; and a kite, made with two cross sticks and a silk handkerchief, enabled the chemical Prometheus to rob the thunder-cloud of its lightnings; but that a worn-out instrument, such as has been just described, should have furnished him who was born to revolutionize the science of the age, with the only means of enquiry at that time within his reach, affords, it must be admitted, a very whimsical illustration of our maxim.

Nor can we pass over these circumstances, without observing how materially they must have influenced the subsequent success of Davy as an experimentalist. Had he, at the commencement of his career, been furnished with all those appliances which he enjoyed at a later period, it is more than probable that he might never have acquired that wonderful tact of manipulation, that ability of suggesting expedients, and of contriving apparatus so as to meet and surmount the difficulties which must ever beset the philosopher in the unbeaten tracks of Science. In this art, he certainly stands unrivalled, and, like his prototype Scheele,* or that pioneer of pneumatic

* Bergman, Professor of Upsal, was informed of a young man who resided in the house of an apothecary, and who was reproached for neglecting the duties of his profession, while he devoted the whole of his time to Chemistry. Bergman's curiosity was excited; he paid him a visit, and was astonished at the knowledge he dis-

experimentalists, Dr. Priestley,* he was unquestionably indebted for his address to the circumstances above related. There never, perhaps, was a more striking exemplification of the adage, that "necessity is the parent of invention."

It would however appear that, imperfect as must have been his apparatus, and limited as were his resources, his ambition very early led him to the investigation of the most abstruse and recondite phenomena. He was not more than seventeen when he formed a strong opinion adverse to the general belief in the existence of *caloric*, or the materiality of heat.

As I shall hereafter have occasion to draw a parallel between the intellectual qualities of Davy, and those of the celebrated Dr. Black, the father of modern chemistry, it may not be irrelevant to state, in this place, that the subject of heat was also amongst the first that attracted the attention of this latter philosopher; indeed, he tells us himself, that he

played, and at the profound researches in which he was engaged, notwithstanding the poverty under which he laboured, and the restraint under which his situation placed him. He encouraged his ardour, and made him his friend. This young man was the celebrated *Scheele*.

* No man ever entered upon an undertaking with less apparent means of success, than did Priestley upon that of Chemistry. He neither possessed apparatus, nor the money to procure it. These circumstances, which at first sight seem so adverse, were in reality those which contributed to his ultimate success. The branch of Chemistry he selected was new; an apparatus had to be invented before any important step could be taken; and as simplicity is essential in every research, he was likely to contrive the best whose circumstances obliged him to attend to economy.

"can scarcely remember the time, when he had not some idea of the disagreement of facts with the commonly received doctrines upon this subject." The tendency of his mind, however, was in direct opposition to that of Davy's, for he insisted upon the materiality of heat, and was the first to conceive the bold idea of its being capable, like any other substance, of entering into chemical combination with various bodies, and of thus losing its characteristic qualities.

Black's theory could not be more opposed to that of Davy than was his conduct upon the occasion; for, although an experiment suggested itself to his mind, by which, as he thought, he could at once establish the truth of his favourite doctrine, he delayed performing it, because there did not happen to be an ice-house in the town in which he lived. With Davy, on the other hand, the conception and execution of an experiment were nearly simultaneous: no sooner, therefore, had he formed his opinion, than his eager spirit urged him to put it to the test.

Having procured a piece of clock-work, so contrived as to be set to work in an exhausted receiver, he added two horizontal plates of brass; the upper one, carrying a small metallic cup to be filled with ice, revolved in contact with the lower one. The whole machine, resting on a plate of ice, was covered by a glass receiver, and the air was exhausted by the very syringe, ingeniously modified for the purpose, with which the reader has already been made acquainted: for, as yet, he had no air-pump, and, what is still more worthy of notice, had never even seen one! The machine was now set in motion, when the ice in the small cup was soon observed to melt; whence he inferred that this effect could alone proceed from vibratory motion, since the whole apparatus was insulated from all accession of material heat, by the frozen mass below, and by the vacuum around it.

The experiment was afterwards repeated with greater care, and by means of a more refined apparatus: it was modified in different ways; and the results were ultimately published in an Essay, to be hereafter noticed, "On Heat, Light, and the Combinations of Light," which appeared in a provincial collection of tracts, edited by Dr. Beddoes, at Bristol.

Mr. Davies Gilbert, in describing the above experiment in his late address to the Royal Society, very justly observed that it does not at all decide the important matter in dispute, with respect to an ethereal or transcendental fluid; but that few young men remote from the society of persons conversant with science, will present themselves, who are capable of devising any thing so ingenious.

Dr. Henry, in a paper published in the "Memoirs of the Manchester Society," on entering into a review of this and similar experiments, very truly states, that the mode of insulation is not only imperfect, but that, according to Count Rumford, caloric will even pass through a Torricellian vacuum.

The most prominent circumstance in the history of this period of Davy's life, is his introduction to Mr. Davies Giddy, now Mr. Gilbert, the late distinguished and popular President of the Royal

Society. The manner in which this happened is as curious as its result was important; and it furnishes another very striking illustration of the power of simple accident in directing our destinies. Mr. Gilbert's attention was attracted to the future philosopher, as he was carelessly swinging over the hatch, or half gate, of Mr. Borlase's house, by the humorous contortions into which he threw his features. Davy, it may be remarked, when a boy, possessed a countenance which, even in its natural state, was very far from comely, while his round shoulders, inharmonious voice, and insignificant manner, were calculated to produce any thing rather than a favourable impression: in riper years, he was what might be called "good-looking," although, as a wit of the day observed, his aspect was certainly of the "Bucolic" character. The change which his person underwent, after his promotion to the Royal Institution, was so rapid, that, in the days of Herodotus, it would have been attributed to nothing less than the miraculous interposition of the Priestesses of Helen. A person, who happened to be walking with Mr. Gilbert upon the occasion alluded to, observed that the extraordinary-looking boy in question was young Davy, the Carver's son, who, he added, was said to be fond of making chemical experiments. "Chemical experiments!" exclaimed Mr. Gilbert, with much surprise: "if that be the case, I must have some conversation with him." Mr. Gilbert, as we all know, possesses a strong perception of character, and he therefore soon discovered ample evidence of the boy's singular genius. After several interviews, which confirmed him in the opinion he had formed, he offered young Humphry the use of his library, or any other assistance that he might require for the pursuit of his studies; and at the same time gave him an invitation to his house at Tredrea, of which he frequently availed himself.

During one of his visits, Mr. Gilbert accompanied him to Hayle Copper-House, and introduced him to Dr. Edwards, a gentleman afterwards known to the medical profession as the chemical lecturer in the school of St. Bartholomew's Hospital; at the time, however, alluded to, he resided at Copper-House with his father, and possessed a well-appointed laboratory. The tumultuous delight which Davy expressed on seeing, for the first time, a quantity of chemical apparatus, hitherto only known to him through the medium of engravings, is described by Mr. Gilbert as surpassing all description. The air-pump more especially fixed his attention, and he worked its piston, exhausted the receiver, and opened its valves, with the simplicity and joy of a child engaged in the examination of a new and favourite toy.

It is a curious circumstance, that the phenomena resulting from the contact of iron and copper, in the investigation of which Davy was destined to perform so prominent a part, were very early noticed by Mr. Edwards in this place; who found that the flood-gates in the Port of Hayle decayed with a rapidity wholly inexplicable, but upon the supposition of some *chemical* action between the metals which had not yet been clearly explained. How little did Mr. Edwards imagine that the fact, which had so powerfully excited his curiosity, would become to the youth before him, a future source of rich and honourable discovery!

During the following year, an event occurred which contributed, in no small degree, to the advancement of Davy's prospects. Mr. Gregory Watt, who had long been in a declining state of health, was recommended by his physicians to reside for some time in the West of England, and he accordingly proceeded at once to Penzance, and took up his abode, as a lodger and boarder, in the house of Mrs. Davy. It may be supposed that two kindred spirits would not be long in contracting an acquaintance with each other; in fact, an intimacy of the warmest nature did ultimately grow up between them, and continue to the very moment of Mr. Watt's premature dissolution: the origin and progress of their friendship, however, are too curious to be passed over without some notice.

Mr. Gregory Watt possessed a warm and affectionate heart; but there was a solemn, aristocratic coldness in his manner, which repulsed every approach to familiarity. Davy, it has been already stated, did not at that time possess any of those qualifications, in person or manner, which are calculated to produce favourable prepossessions. It may, therefore, be readily imagined how Mr. Watt must have felt, on finding the son of his landlady familiarly addressing him on subjects of metaphysics and poetry. By one of those strange perversions which have so frequently led great men to conceal the peculiarity of their talents, and to rest their claims to notice and respect upon qualifications which they possessed only in an inferior degree, Davy sought to ingratiate himself with Mr. Watt by metaphysical discussions; but, instead of the admiration, he excited the disgust of his hearer. It was by mere accident that an allusion was first made to chemistry, when Davy flippantly observed, that he would undertake to demolish the French theory in half an hour. He had touched the chord: the interest of Mr. Watt was excited,—he conversed with Davy upon his chemical pursuits,—he was at once astonished and delighted at his sagacity,—the barrier of ice was removed, and they became attached friends.

Mr. Wedgwood, and his brother Thomas, also spent a winter at Penzance; and I have reason to believe that their friendship was of substantial benefit to Davy.

Before I attend the progress of our philosopher to the next scene of life, or proceed to detail the circumstances connected with his departure from Penzance, I must relate the following anecdote. —Until the formation of the Geological Society of London occasioned the introduction of more extended and sounder views into the science, geologists were divided into two great rival sects,—into Neptunists and Plutonists: the one affirming that the globe was exclusively indebted for its present form and arrangement to the agency of water; the other, admitting to a certain extent the operation of water, but maintaining the utter impossibility of explaining the consolidation of the strata without the intervention of fire. Every geologist felt bound to side with the one or the other of these contending parties, for neutrality was held as disgraceful as though the law of Solon had been in active operation. I shall not easily forget the din and fury of this elemental war, as it raged in Edinburgh when I was a student in that University; even the mineral dealers, who, like the artisans of a neutral city, sold arms and ammunition to both sides, still defended their own opinions with party fury. It was amusing to observe the triumph and dismay which, by turns, animated and depressed each side, as the discovery of a new fact, or a fresh specimen, appeared to give a preponderance to the doctrine of fire or water. The fact of so large a portion of the strata being found in the state of a carbonate was advanced by the Neptunists as an unanswerable argument against igneous agency: the dismay therefore which this sect received upon the discovery of Sir James Hall, that under the combined forces of heat and compression, carbonate of lime might be fused, was only equalled by the excessive joy excited in the contending party. We may form some notion of the high importance attached to this discovery, when we learn that its author applied to the Government for a flag of truce to convey illustrative specimens to the Continental philosophers.

It so happened, that the Professors of Oxford and Cambridge ranged themselves under opposite banners: Dr. Beddoes was a violent and uncompromising Plutonist, while Professor Hailstone was as decided a Neptunist. The rocks of Cornwall, and their granitic veins, had been appealed to, as affording evidence upon the subject; and the two Professors, who, although adverse in opinion, were united in friendship, determined to proceed together to the field of dispute, each hoping that he might thus convince the other of his error, and cure him of his heresy. The belligerents arrived at Penzance, and in company with their mutual friend, Mr. Davies Gilbert, examined the coast, and procured specimens with pretty much the same spirit of selection as a schoolboy consults his Gradus, not for an epithet of any meaning, but for one which best suits his measure; and having made drawings, disputed obvious appearances, rendered that which was clear to the senses, confused to the understanding, and what was already confused, ten times more obscure, they returned, the opinion of each, as might easily have been anticipated, having been strengthened by the ordeal: the one protesting that the very aspect of the shivered slate was sufficient to prove that the globe must have been roasted to rags; the other, with equal plausibility, declaring that there was not a tittle of evidence to show that the watery solvent had ever even simmered. Such, in fact, must ever be the case, when philosophers examine the same subject under such different impressions, and in such opposite points of view; like the two knights who could not agree respecting the colour of the shield, only because each saw a different side of it.

Rocks, it is said, have flinty hearts, and certain it is that, upon this occasion, Cornwall did not afford that assistance against the Neptunists, which the Oxford Professor had sought with so much zeal and confidence; but if deferred revenge had, as we are told is generally the case, been put out at compound interest, and Beddoes had exacted its dues with more than judaical rigour, it must be allowed that Cornwall, by placing Davy at his disposal, would have fully cancelled all demands.

> Plutonian Beddoes, erst, in spiteful ire,
> To see a *Hailstone* mock his central fire,
> A mighty spirit raised, by whose device
> We now burn HAILSTONES, and set fire to ICE.

Before quitting this subject, it is but justice to advert to the progress which Geology has made since the turbulence of this contest has subsided; it has grown strong in facts, and is daily increasing its stores. It has been wisely said by one of the ancient Poets, that in vehement disputes, not only the persons engaged, but every one who is at all interested, must suffer; not only the combatants, but the spectators of the combat,—for it is difficult to apprehend truth while it is the subject of angry contest.

To return to the narrative.—Upon Beddoes establishing the "Pneumatic Institution" at Bristol, he required an assistant who might superintend the necessary experiments in the laboratory; and Mr. Gilbert proposed Davy as a person fully competent to fill the situation. The young candidate had already produced a very favourable impression upon Dr. Beddoes, by his experiments upon Heat and Light, which he had some time before transmitted to him through the hands of his friend Mr. Gregory Watt. This fact may be collected from a note appended by Dr. Beddoes to Davy's paper subsequently published in the first volume of the West Country Contributions, in which the Doctor says,

"My first knowledge of Mr. Davy arose from a letter written in April 1798, containing an account of his researches on Heat and Light." The rest is told in the letters which passed on this occasion between Dr. Beddoes and Mr. Gilbert, and from which I shall make such extracts as may be necessary to complete the history of a transaction of much interest and importance.

In a letter dated July 4, 1798, Dr. Beddoes says, I am glad that Mr. Davy has impressed you as he has me. I have long wished to write to you about him, for I think I can open a more fruitful field of investigation than any body else. Is it not also his most direct road to fortune? Should he not bring out a favourable result, he may still exhibit talents for investigation, and entitle himself to public confidence more effectually than by any other mode. He must be maintained, but the fund will not furnish a salary from which a man can lay up any thing. He must also devote his time for two or three years to the investigation. I wish you would converse with him upon the subject. No doubt he has received my two last letters. I am sorry I cannot at this moment specify a yearly sum, nor can I say with certainty whether all the subscribers will accede to my plan; most of them will, I doubt not. I have written to the principal ones, and will lose no time in sounding them all."

In a second letter of the 18th of July, we find the following observations. "I have received a letter from Mr. Davy since I wrote to you. He has oftener than once mentioned a *genteel maintenance*, as a preliminary to his being employed to superintend the Pneumatic Hospital. I fear the funds will not allow an ample salary; he must, however, be maintained. I can attach no idea to the epithet *genteel*, but perhaps all difficulties would vanish in conversation; at least, I think your conversing with Mr. Davy will be a more likely way of smoothing difficulties, than our correspondence. It appears to me, that this appointment will bear to be considered as a part of Mr. Davy's medical education, and that it will be a great saving of expense to him. It may also be the foundation of a lucrative reputation; and certainly nothing on my part shall be wanting to secure to him the credit he may deserve. He does not undertake to discover cures for this or that disease; he may acquire just applause by bringing out clear, though negative results. During my journeys into the country, I have picked up a variety of important and curious facts from different practitioners. This has suggested to me the idea of collecting and publishing such facts as this part of the country will, from time to time, afford. If I could procure chemical experiments, that bore any relation to organised nature, I would insert them. If Mr. Davy does not dislike this method of publishing his experiments, I would gladly place them at the head of my first volume, but I wish not that he should make any sacrifice of judgment or inclination."

It remains only to be stated, that Mr. Gilbert kindly undertook the negotiation, and completed it to the satisfaction of all the principal parties. Mrs. Davy yielded to her son's wishes, and Mr. Borlase very generously surrendered his indenture, with an endorsement to the following effect, — that he freely gave up the indenture, on account of the singularly promising talents which Mr. Davy had displayed.

His old and valued friend Mr. Tonkin, however, not only expressed his disapprobation of this scheme, but was so vexed and irritated at having his favourite plan of fixing Davy in his native town as a Surgeon, thus thwarted, that he actually altered his will, and revoked the legacy of his house which he had previously bequeathed him. Mr. Tonkin died on the 24th of December 1801; so that, although he lived long enough to witness Davy's appointment to the Royal Institution, he could never have anticipated the elevation to which his genius and talents ultimately raised him.

On the 2nd of October in the year 1798, Davy quitted Penzance, before he had attained his twentieth year. Mr. Gilbert well remembers meeting him upon his journey to Bristol, and breakfasting with him at Okehampton, on the 4th of October. He was in the highest spirits, and in that frame of mind in which a man of ardent imagination identifies every successful occurrence with his own fortunes; his exhilaration, therefore, was not a little heightened by the arrival of the mail-coach from London, covered with laurels and ribbons, and bringing the news, so cheering to every English heart, of NELSON's glorious victory of the NILE.

CHAPTER II.

Cursory thoughts on the advantages of Biography.—Plan and objects of the Pneumatic Institution.—Davy contracts friendships during his residence at Bristol.—His first visit to London.—His Letters to Mr. Davies Gilbert.—The publication of the West Country Contributions, by Dr. Beddoes.—Davy's Essays on Heat, Light, and Respiration.—His interesting experiments on bonnet canes.—He commences an enquiry into the nature of nitrous oxyd.—He publishes his chemical researches.—A critical examination of the work.—Testimony of Tobin, Clayfield, Southey, and others, respecting the powers of nitrous oxyd.—Davy breathes carburetted hydrogen gas, and nearly perishes from its effects.—His new Galvanic experiments communicated in a Letter to Mr. Gilbert.

Having concluded the early history of the subject of these memoirs, and conducted it to that memorable day on which he left his native town, and bursting from obscurity, prepared to enter upon a wider field of usefulness and honour, I shall accompany him in his progress; and with the honest desire of affording instruction as well as amusement, — for history is useful only as it holds up the mirror of Truth, — I shall continue to point out the various circumstances that may have contributed to his success and scientific renown; and to offer such occasional reflections as may be likely to illustrate not only the superficial peculiarities which constitute the light and shade of character, but those deeper varieties of mind, upon which the superiority of intellect may be supposed to depend.

After all, the great end of biography is not to be found, as some would seem to imagine, in a series of dates, or in a collection of gossiping anecdotes and table-talk, which, instead of lighting up and vivifying the features, hang as a cloud of dust upon the portrait; but it is to be found in an analysis of human genius, and in the developement of those elements of the mind, to whose varied combinations, and nicely adjusted proportions, the mental habits, and intellectual peculiarities of distinguished men may be readily referred.

It has been stated that an arrangement had been concluded between Dr. Beddoes and Davy: it is but an act of justice to say, that it was of a liberal and honourable description; and let me also add in this place, that no sooner had Davy found himself in a situation which secured for him the necessaries of life, than he renounced all claims upon his paternal property, in favour of his mother and sisters.

By acceding to the proposal of Dr. Beddoes, he never intended to abandon the profession in which he had embarked; on the contrary, he persevered in his determination to study and graduate at Edinburgh, and his patron promised that every opportunity should be afforded him at Bristol for seeing medical practice: this part of the arrangement, however, was voluntarily abandoned by him, for he soon became so absorbed by the labours of the laboratory, as to leave little leisure for the clinical studies of the hospital.

The Pneumatic Institution was established for the purpose of investigating the medical powers of factitious airs or gases; and to Davy was assigned the office of superintending the various experiments.

It is now generally acknowledged, that the Art of Physic has not derived any direct advantage from the application of a class of agents which, undoubtedly, held forth the fairest promise of benefit; but it is too frequently the case, that in physic, theory and experience are in open hostilities with each other. The gases are now never employed in the treatment of disease, except by a few crafty or ignorant empirics, whose business it is to enrich themselves by playing on the credulity of mankind: indeed, we may say of popular remedies in general what M. de Lagrange has so wittily said of popular prejudices, that they are the cast-off clothes of philosophers, in which the rabble dress themselves.

The investigation, however, into the nature and composition of the gases paved the way to some new and important discoveries in science; so that, to borrow a Baconian metaphor, although our philosophers failed in obtaining the treasure for which they so eagerly dug, they at least, by turning up and pulverizing the soil, rendered it fertile. The ingenuity of the chemist will for ever remain on record; the phantoms of the physicians have vanished into air.

Davy was now constantly engaged in the prosecution of new experiments, in the conception of which, as he himself informs us, he was greatly aided by the conversation and advice of Dr. Beddoes. He was also occasionally assisted by Mr. William Clayfield, a gentleman ardently attached to chemical pursuits, and whose name is not unknown in the annals of science; indeed, it appears that to him he was indebted for the invention of a mercurial air-holder, by which he was enabled to collect and measure the various gases submitted to examination. He had also the advantages of some society of a highly intellectual cast: it is sufficient to mention the names of Edgeworth and James Tobin.

In reply to a letter of enquiry which was lately addressed to her, Miss Edgeworth observes, that "her father possessed much influence over Davy's mind;" and that "when he was a very young man at Clifton, unknown to fame, Mr. Edgeworth early distinguished and warmly admired his talents, and gave him much counsel, which sunk deep into his mind."

The present Lord Durham and his brother were also resident in the house of Dr. Beddoes, not only for their education, but for the benefit of his professional superintendence. Besides those who were residing at Clifton, the most distinguished in the circles of science and literature paid passing visits to Dr. Beddoes; with many of whom Davy contracted an acquaintance, with some an intimacy, and with a few a solid and permanent friendship. In examining the individuals composing this latter class, we find them differing so widely from each other in character and pursuit, that we are led to enquire upon what principles of affinity his regards could possibly have been attracted — the truth is, that there was more than one avenue to his heart; and the philosopher, the poet, the physician, the philanthropist, and the sportsman, found each, upon different terms, a more or less ready access to its recesses. The chemist who would aspire to his favour, could alone obtain it by laborious application and novel research; the philanthropist, by the practicability of his schemes for improving society, and increasing the sum of its happiness; but the fisherman instantly caught his affections by a hook and line. To be a fly-fisher was, in his opinion, to possess the capabilities of intellectual distinction, although circumstances might not have conspired to call them into action; whilst a proficiency in this art, when exhibited by an individual otherwise distinguished, gave him an additional claim to his attention and regard. The stern courage of Nelson, tempered as it was with all the kindly feelings of humanity, was sufficient to excite in the breast of Davy the most enthusiastic admiration; but the circumstance of his having been a fly-fisher, and continued the sport, even with his left hand, threw, in his opinion, a still brighter halo around his character.

No one who knew him can accuse him of inconstancy in his friendships: amidst the excitements of his station, and the abstractions incident to his pursuits, he might not always have shown those little attentions which are received by the world as the indications of personal regard; but his heart beat not less warmly on that account: when the flame of affection had been once kindled, it burnt with a pure and steady light through life. This will be readily seen in the letters addressed to his several early friends, more especially to Mr. Poole of Nether Stowey, in Somersetshire, and to Mr. Clayfield of Bristol, from which I shall have occasion to present some interesting extracts.

Those who had become acquainted with him in early life, and were enabled to watch the whole progress of his career from obscurity to the highest pinnacle of fame, have declared that his extraordinary talents never at any period excited greater astonishment and admiration than during his short residence at Bristol. His simplicity of mind and manner was also at this time truly delightful. He scarcely knew the names of our best authors, much less read any of their works; yet upon topics of moral philosophy and metaphysics he would enter into discussion with acknowledged scholars, and not only delight them with the native energy of his mind, but instruct them by the novelty and truth of his conceptions. Mr. Coleridge lately expressed to me the astonishment he felt, very shortly after his introduction to him, on hearing him maintain an argument upon some abstruse subject with a gentleman equally distinguished for the extent of his erudition, and for the talent of rendering it available for illustration;—the contrast was most striking—it was the fresh and native wild flower, opposed to the elaborate exotic of the *Hortus Siccus!*

During this period he occasionally visited his friend Mr. Gregory Watt, at Birmingham; at which place his ambition was constantly excited by intercourse with congenial minds; and his letters to his mother and relations represent him as rejoicing in the success of his experiments, and as delighting in his association with kindred genius; but always casting a longing, lingering thought on the scenes of his boyhood, he spoke with joyful anticipation of the period at which he proposed to revisit his mother and family.

That he still continued to regard the practice of physic as the great end and object of all his pursuits, is evident from one of these letters, written in 1799, in which he says, "Philosophy, Chemistry, and Medicine, are my profession."

On the 1st of December 1799 he visited London for the first time, and remained about a fortnight; the friends with whom he associated upon this occasion were Coleridge, Southey, Gregory Watt, Underwood, James and John Tobin, Thomson, and Clayfield; all of whom vied with each other in their exertions to render his visit agreeable, conducting him to such persons and places as were deemed worthy of his notice.

Of all the letters placed at my disposal, those addressed to his early friend and patron, Mr. Davies Gilbert, are, in my judgment, the most interesting: it is true, that as specimens of epistolary style they have but slender pretensions, and are far less pleasing than those written to Mr. Poole and others, in later life; but let it be remembered that, as yet, their writer had never enjoyed the advantages of literary correspondence. For the defects, however, of style, there is more than sufficient compensation; they speak from the heart;—they carry with them internal evidence of the honest simplicity of his mind, and they throw a light upon the peculiarities of his genius, which without such aid might be less perfectly understood; above all, they evince an ardour which no difficulties could repress, and a confidence which no failures could extinguish. We clearly discern from his first letters, that he entered upon his career of experiment with an almost chivalrous feeling, flushed with the consciousness of native strength, and exulting in the prospect of destined achievements.

I am aware that there are those who still object, with Dr. Sprat, to the practice of publishing letters which were never intended for the public eye, and I experience the inconvenience, while I respect the delicacy, of such an opinion. I confess, on my own part, I have always considered, with Mr. Mason, that the objections urged by the learned historian of the Royal Society are wholly untenable. He talks of "the souls of men thus appearing undressed, or in a habit too negligent to go abroad in the streets, although they might be seen by a few in a chamber." But the undress he would condemn, is the nakedness of Truth—the negligent attire, the simple and unadorned expression of those natural and significant traits, whose value incomparably exceeds the premeditated and artificial exhibitions of mind and manner. "*Nam in ingenio quoque sicut in agro, quanquam alia diu serantur atque elaborantur, gratiora tamen quæ sua sponte nascuntur.*"*

* Dialogus de Oratoribus,—*Tacit.*

I cannot but suspect that Dr. Sprat was, upon this occasion, more anxious to display a metaphor, than to illustrate a truth. I have often thought a very curious book might be written to show how greatly, both in physics and in morals, the progress of truth has been retarded, and the judgment of men warped, by the abuse of metaphors; the most correct of which can be nothing more than the image of Truth reflected, as it were, from a mirror, and consequently liable to all the delusions of our mental optics. The figure by which Nature was represented as "*abhorring a vacuum*," kept us in ignorance of the true theory of the pump for two thousand years after the discovery of the weight, or gravity, of the atmosphere;* and the unfortunate P. L. Courier positively owed his conviction to a metaphor in the Judge's charge—"*Un écrit plein de poison.*"—Well might the defendant exclaim, "*Sauvez-nous de la Metaphore!*"

The first of the letters to which I have alluded appears to have been written rather more than five weeks after his arrival at Clifton.

TO DAVIES GIDDY, ESQ.

DEAR SIR, Clifton, November 12, 1798.

I HAVE purposely delayed writing until I could communicate to you some intelligence of importance

* Plutarch, in expressing the opinion of Asclepiades upon this subject, represents him as saying, that the external air, *by its weight*, opened its way with force into the breast. Seneca also was acquainted with the weight and elastic force of the air; for he describes the constant effort by which it expands itself when it is compressed, and affirms that it has the property of condensing itself, and of forcing its way through all obstacles that oppose its passage.—Quæst. Nat. lib. v. c. v. and vi.

concerning the Pneumatic Institution. The speedy execution of the plan will, I think, interest you, both as a subscriber and a friend to science and mankind. The present subscription is, we suppose, nearly adequate to the purpose of investigating the medicinal powers of factitious airs; it still continues to increase, and we may hope for the ability of pursuing the investigation to its full extent. We are negotiating for a house in Dowrie Square, the proximity of which to Bristol, and its general situation and advantages, render it very suitable to the purpose. The funds will, I suppose, enable us to provide for eight or ten patients in the hospital, and for as many out of it as we can procure.

We shall try the gases in every possible way. They may be condensed by pressure and rarefied by heat. *Quere*,—would not a powerful injecting syringe,* furnished with two valves, one opening into an air-holder and the other into the breathing chamber, answer the purpose of compression better than any other apparatus? Can you not, from your extensive stores of philosophy, furnish us with some hints on this subject? May not the non-respirable gases furnish a class of different stimuli? of which the *oxy-muriatic acid gas* would stand the highest, if we might judge from its effects on the lungs; then, probably, *gaseous oxyd of azote*, and *hydrocarbonate*.

I suppose you have not heard of the discovery of the native *sulphate of strontian* in England. I shall perhaps surprise you by stating that we have it in large quantities here. It had long been mistaken for

* Here the reader will recognise the force of early associations.

sulphate of barytes, till our friend Clayfield, on endeavouring to procure the *muriate of barytes* from it by decomposition, detected the strontian. We opened a fine vein of it about a fortnight ago, at the Old Passage near the mouth of the Severn. It was embodied in limestone and gypsum, the outside of the vein, a striated mass; the internal parts finely crystallised in cubes, of the Sp. Gr. 4·1. Clayfield has been working at it for some time. We have persuaded him to publish his analysis in the first volume of the Western Physical Collection.

I have made with him the phosphuret of barytes and of strontian: they possess, in common with that of lime, the property of producing phosphorized hydrogen gas; the phosphuret of strontian, it appears, in a more eminent degree.

We have likewise attempted to decompose the boracic and muriatic acids, by passing phosphorus, in vapour, through muriate, and borate of lime, heated red. Phosphate of lime was found in the experiment on the boracic acid; but, as no pneumatic apparatus was employed, the experiment was uncertain. We shall repeat them next week.

We are printing in Bristol the first volume of the 'West Country Collection,' which will, I suppose, be out in the beginning of January.

Mrs. Beddoes hopes that Miss Giddy received her letter, and desires me to certify that she wrote almost immediately after the reception of her epistle. She is as good, amiable, and elegant as when you saw her. Believe me, dear Sir, with affection and respect,

Truly your's,

HUMPHRY DAVY.

The work announced in the above letter was pub lished in the commencement of the year 1799, under the title of "Contributions to Physical and Medical Knowledge, principally from the West of England; collected by Thomas Beddoes, M.D."

The first two hundred pages, constituting very nearly half the volume, are the composition of Davy, and consist of essays "On Heat, Light, and the Combinations of Light." "On Phos-oxygen, or Oxygen and its Combinations;" and "On the Theory of Respiration."

His first essay commences with an experiment, in order to show that light is not, as Lavoisier supposed, a modification, or an effect, of heat, but matter of a peculiar kind, *sui generis*, which, when moving through space, or in a state of projection, is capable of becoming the source of a numerous class of our sensations.

A small gunlock was armed with an excellent flint, and, on being snapped in an exhausted receiver, did not produce any light. The experiment was repeated in carbonic acid, and with a similar result. Small particles were in each case separated from the steel, which, on microscopic examination, evidently appeared to have undergone fusion. Whence Davy argued, that light cannot be caloric in a state of projection, or it must have been produced in these experiments, where heat existed to an extent sufficient to fuse steel. Nor, that it can be, as some have supposed, a vibration of the imaginary fluid ether; for, granting the existence of such a fluid, it must have been present in the receiver. If, then, light be neither caloric in a state of projection, nor the vibration of an imaginary ether, it must, he says, be a substance *sui generis*.

With regard to caloric, his opinion that it is not, like light, material, has been already noticed. In the present essay he maintains the proposition by the same method of reasoning as that by which he attempts to establish the materiality of light, and which mathematicians have termed the "*reductio ad absurdum*."

In his chapter on "Light and its Combinations," he indulges in speculations of the wildest nature, although it must be confessed that he has infused an interest into them which might almost be called dramatic. They are certainly highly characteristic of that enlightened fancy, which was perpetually on the wing, and whose flight, when afterwards tempered and directed by judgment, enabled him to abstract the richest treasures from the recesses of abstract truth.

Taking it for granted that caloric has no existence as a material body, or, in other words, that the phenomena of repulsion do not depend upon the agency of a peculiar fluid, and that, on the contrary, light is a subtle fluid acting on our organs of vision *only when in a state of repulsive projection*, he proceeds to examine the French theory of combustion; the defects of which he considers to arise from the assumption of the imaginary fluid *caloric*, and the total neglect of *light*. He conceives that the light evolved during combustion previously existed in the oxygen gas, which he therefore proposes for the future to call PHOS-OXYGEN.*

* Brugnatelli considered that oxygen, in certain cases of combination, entered into union with different bodies, without parting

In following up this question, he would seem to consider Light as the *Anima Mundi*, diffusing through the universe not only organization, but even animation and perception.

Phos-oxygen he considers as capable of combining with additional proportions of Light, and of thus becoming '*luminated Phos-oxygen !*'—from the decomposition of which, and the consequent liberation of light, he seeks to explain many of the most recondite phenomena of Nature.

We cannot but admire the eagerness with which he enlists known facts into his service, and the boldness with which he ranges the wilds of creation in search of analogies for the support and illustration of his views. He imagines that the *Phos-oxygen*, when thus *luminated*, must necessarily have its specific gravity considerably diminished by the combination, and that it will therefore occupy the higher regions of the atmosphere; hence, he says, it is that combustion takes place at the tops of mountains at a lower temperature than in the plains, and with a greater liberation of light. The hydrogen which is disengaged from the surface of the earth, he supposes, will rise until it comes into contact with this *luminated Phos-oxygen*, when, by its attracting the oxygen to form water, the light will be set free, and give origin to the phenomena of fiery meteors at a great altitude.

The phenomenon termed '*Phosphorescence*,' or that luminous appearance which certain bodies exhibit after exposure to heat, is attributed by this

with its *caloric;* and in that state he gave it the name of THERMOXYGEN; so that Davy had a precedent for his nomenclatural innovation.

theory to the light, which may be supposed to quit such substances as soon as its particles have acquired repulsive motion by elevation of temperature.

The Electric Fluid is considered as Light in a condensed state, or, in other words, in that peculiar state in which it is not supplied with a repulsive motion sufficiently energetic to impart projection to its particles; for, he observes, that its chemical action upon bodies is similar to that of Light; and when supplied with repulsive motion by friction, or by the contact of bodies from which it is capable of subtracting it, it loses the projectile form, and becomes perceptible as Light. It is extremely probable, he adds, that the great quantity of this fluid almost everywhere diffused over our earth is produced by the condensation of Light, in consequence of the subtraction of its repulsive motion by black and dark bodies; while it may again recover the projectile force by the repulsive motion of the poles, caused by the revolution of the earth on its axis, and thus appear again in the state of sensible light; and hence the phenomenon of the *Aurora Borealis*, or Northern Lights.

In considering the theory of Respiration, he supposes that *phos-oxygen* combines with the venous blood without decomposition; but that, on reaching the brain, the light is liberated in the form of Electricity, which he believes to be identical with the nervous fluid. On this supposition, sensations and ideas are nothing more than motions of the nervous ether; or light exciting the medullary substance of the nerves and brain into sensitive action!

He thinks it would be worth while to try, by a

very sensible electrometer, whether an insulated muscle, when stimulated into action, would not give indications of the liberation of electric fluid, although he suspects that in man the quantity is probably too small, and too slowly liberated, to be ascertainable. In the torpedo, and in some other animals, however, it is unquestionably given out perceptibly during animal action.

When any considerable change takes place in the organic matter of the body, so as to destroy the powers of life, new chemical attractions and repulsive motions take place, and the different principles of which the body is composed enter into new combinations. In this process, which is called putrefaction, Davy, in pursuance of this theory, thinks that in land-animals the latent light of the system enters into new combinations with oxygen and nitrogen, but that in fish no such combinations occur, and hence the luminous appearance which accompanies their putrefaction.

Such is the outline of these extraordinary Essays. They stand upon record, and therefore, as a faithful biographer, I was bound to notice them; nor are they devoid of interest or instruction: I am not quite sure that, amidst all the meteors of his fancy, there may not be a gleam of truth. I allude to his theory of Respiration: it certainly does not square with the physiological opinions of the day; nor did that of Newton, when he conjectured that water might contain an inflammable element; but it was the refraction of a great truth, at that time below the horizon.

It was a very ancient opinion, that life, being in

its own nature aëriform, is under the necessity of renewing itself by inspiring the air. Modern chemistry, by teaching us the nature of the atmosphere, has dispelled many fanciful theories of its action, but it has not yet explained why respiration, the first and last act of life,* cannot be suspended, even for a minute, without the extinction of vitality. When we reflect upon this fact, it is scarcely possible not to believe that the function has been ordained for some greater purpose than that of removing a portion of carbon from the circulating blood. Is it unreasonable to conclude that some principle is thus imparted, which is too subtle to be long retained in our vessels, and too important to be dispensed with, even for the shortest period? "I offer this opinion," as Montaigne says, "not as being good, but as being my own."

By these observations, I am not to be supposed as wishing, for a moment, to uphold the wild hypotheses which I have just related; it must be admitted that the theory of *phos-oxygen* and *luminated phos-oxygen* has scarcely a parallel in extravagance and absurdity; and I happen to know that, in after life, Davy bitterly regretted that he had so committed himself; any allusion to the subject became a source of painful irritation. It is to be remarked, that in every course of lectures, although Davy did not refer to these theories, he frequently alluded to the

* Breath and life are synonymous. In the Greek, the most philosophically constructed language with which we are acquainted, this *first* and *last* act is expressed by a verb composed of alpha and omega—*αω*. In the Latin, the connexion between *spiro* and *spiritus*, breath and life, is evident.

unphilosophic spirit that had given origin to them; as if he had imposed upon himself this penance as an atonement for his early follies. The following note was taken at one of his lectures:—"After what has been said, it will be useless to enter upon an examination of any of those theories, which, assuming for their foundation the connexion of life with respiration, have attempted to prove that oxygen is the principle of life, and that the wonderful and mysterious phenomena of perception arise from the action of common gravitating substances upon each other. Such theories are the dreams of misemployed genius, which the light of experiment and observation has never conducted to truth, and are merely a collection of terms derived from known phenomena, and applied by loose analogies of language to unknown things."

The reader, however, will be disposed to treat him with all tenderness when he remembers that the author of these Essays was barely eighteen years of age. If blame is to fall on any one, let it fall on Dr. Beddoes, who never should have sanctioned the publication: had he curbed the ardent and untamed imagination of the young philosopher, he would have acted the part of a wise man and of a kind friend. But the truth is, that much as Davy needed the bridle, Beddoes * required it still more; for, notwithstanding his talents, he was as little fitted for a Mentor as a weathercock for a compass; and had it

* The only pun Davy is said to have ever made was upon the occasion of Mr. Sadler being appointed by Dr. Beddoes as his successor. "I cannot imagine," said he, "why he has engaged *Sadler*, unless it is that he may be well *bridled*."

not been for the ascendency which Davy gained over his mind, the ardour of his temperament would have continually urged him beyond the bounds of reason.

Caught by the loosest analogies, he would arrive at a conclusion without examining all the conditions of his problem. In the exercise of his profession, therefore, he was frequently led to prescribe plans which he felt it necessary to retract the next hour. His friend Mr. T—— had occasion to consult him upon the case of his wife: the Doctor prescribed a new remedy; but, in the course of the day he returned in haste, and begged that, before Mrs. T—— took the medicine, its effect might be tried on a dog!

The following anecdote, which was lately communicated to me by Mr. Coleridge, will not only illustrate a trait of character, but furnish a salutary lesson to the credulous patron of empirics. As soon as the powers of nitrous oxide were discovered, Dr. Beddoes at once concluded that it must necessarily be a specific for paralysis. A patient was selected for the trial, and the management of it was entrusted to Davy. Previous to the administration of the gas, he inserted a small pocket thermometer under the tongue of the patient, as he was accustomed to do upon such occasions, to ascertain the degree of animal temperature, with a view to future comparison. The paralytic man, wholly ignorant of the nature of the process to which he was to submit, but deeply impressed, from the representations of Dr. Beddoes, with the certainty of its success, no sooner felt the thermometer between his teeth than he concluded that the *talisman* was in full operation, and in a burst of enthusiasm declared that he already experienced the effects of its benign influence throughout his whole body:—the opportunity was too tempting to be lost—Davy cast an intelligent glance at Mr. Coleridge, and desired the patient to renew his visit on the following day, when the same ceremony was again performed, and repeated every succeeding day for a fortnight, the patient gradually improving during that period, when he was dismissed as cured, no other application having been used than that of the thermometer. Dr. Beddoes, from whom the circumstances of the case had been intentionally concealed, saw in the restoration of the patient the confirmation of his opinion, and the fulfilment of his most ardent hope —Nitrous Oxide was a specific remedy for Paralysis! "It were criminal to retard the general promulgation of so important a discovery; it were cruel to delay the communication of the fact until the publication of another volume of his '*Contributions;*' the periodical magazines were too slow in their rate of travelling,—a flying pamphlet would be more expeditious; paragraphs in the newspapers; circulars to the hospitals:"—such were the reflections and plans which successively agitated the physician's mind, when his eyes were opened to the unwelcome truth by Davy's confessing the delusion that had been practised.

A short time after the publication of the first volume of the "Contributions," Davy addressed to his friend the following letter:—

TO DAVIES GIDDY, ESQ.

Clifton, Feb. 22, 1799.

Dear Friend,—for I love you too well to call you by a more ceremonious name,—I have delayed writing to you from day to day, expecting that some of our experiments would produce results worthy of communication. Since I received your last very acceptable letter, I have been chiefly employed in pursuing the experiments on Heat, Light, Respiration, &c. Of these experiments I shall give you no account, as you will see them in print. I sent you a copy of my Essays last week; if you have not received them, I trust you will find them at my mother's, or at Mr. Tonkin's.

In the same parcel were two small packets, one from Mrs. Beddoes for your father, the other for Miss Giddy from Mrs. Willoughby. About a fortnight ago I sent a few chemical instruments to Mr. Penneck of Penzance; and inclosed with them were specimens of the different varieties of sulphate of strontian addressed to you. If you have not received them, you will get them by sending to Mr. Penneck.

On looking over a box of minerals last week, which was sent to Dr. Beddoes from Cumberland, I found two very fine specimens of sulphate of strontian, marked by the collector *Laminated Shorl.* I suspect this mineral is not scarce in calcareous countries; it is, I dare say, often mistaken for sulphate of barytes.

I have succeeded in combining strontian with the oxygenated muriatic acid. This salt possesses most astonishing properties;* you will find an account of them in my Essay.

When you have perused my papers, I shall be very much obliged to you for a criticism upon them. When I left Penzance, I was quite an infant in speculation,—I knew very little of Light or Heat. I am now as much convinced of the non-existence of caloric, as I am of the existence of light. Independent of the experiments which appear to demonstrate its non-existence directly, and of which you will find an account in my Essay, the consideration of certain phenomena leads me to suppose that there would be no difficulty in proving its non-existence by reasoning. These considerations have occurred to me since the publication of the work. I could now render it much more perfect; but I hope soon to complete the investigation of the combinations of Light, and to produce a much more perfect work on the subject. I shall be infinitely obliged to you for any hints or observations,

* The following is the account given in his Essay. "When sulphuric acid was poured into a solution of this salt in water, a beautiful and unexpected phenomenon took place. The room was accidentally darkened at the moment this experiment was made, so that we were enabled to perceive a vivid luminous appearance. This experiment, independent of its beauty, is extremely pleasing as affording an instance of true combustion, that is, the production of Light and Heat by the mixture of two incombustible bodies." It may be presumed, that this phenomenon arose from the developement and decomposition of a portion of Euchlorine, a compound which he subsequently discovered in 1811. In the year 1813, Chevreul announced, as a new discovery, that if strontian be heated in contact with muriatic acid gas, the gas is absorbed, and the earthy salt becomes red hot.—*See Annals of Philosophy*, vol. ii. p. 312.

as far as the detection of errors of any kind, for it is no flattery to say that I pay greater deference to your opinion than to that of any other philosopher.

We intend next week to endeavour to ascertain, by the aid of a delicate balance, the quantities of Light liberated in different combustive processes. That there is a deficiency of weight, I am convinced from many experiments.

The experiments on Light, &c. have prevented me from attempting the decomposition of the undecompounded acids. We have ordered an apparatus at the glass-house for this purpose, and I hope next week we shall be able to carry on the investigation. Two modes of effecting these decompositions have occurred to me:—first, to bring phosphorus or sulphur, in the gaseous state, in contact with the acid gases in a tube heated intensely; secondly, to send sulphur in the gaseous state through muriate of copper or lead, heated white. The attraction of sulphur for oxygen, of copper for oxygen, and of sulphur for copper, will probably effect the decomposition.

Our laboratory in the Pneumatic Institution is nearly finished, and we shall begin the investigations in about a fortnight. We shall begin by trying the gases in their simplest mode of application, and gradually carry on the more complex processes.

I hope the gaseous oxide of azote will prove to be a specific stimulus for the absorbents.

I was last week surprised by a letter from Mr. Watt, announcing the success of their trial. When I was at Birmingham five weeks ago, the family were in very low spirits. I spent nine or ten days there, chiefly with Mr. Keir and Mr. Watt: I had a great deal of chemical conversation with them. Mr. Keir is one of the best-informed men I have ever met with, and extremely agreeable. Both he and Mr. Watt are still phlogitians; but Mr. Keir altogether disbelieves the doctrine of *calorique.*

What news have you in Cornwall? Has Mr. John Hawkins returned to his native county? he will doubtless be a great acquisition to you.

Pray do you know whether the Zoophyta and marine worms are susceptible of the galvanic stimulus? Experiments on them would go far to determine whether the irritable or sensitive fibre is primarily affected.

I know of little general scientific news. In the last volume of the *Annales de Chimie* is a curious paper by Berthollet on sulphurated hydrogen: he makes it out to be an acid. I shall most anxiously expect a letter from you, and I remain with affection and respect, Yours,

HUMPHRY DAVY.

The letter which follows may be considered as a reply to one received from Mr. Davies Gilbert, which, it would appear, contained strictures upon his recently published Essays.

TO DAVIES GIDDY, ESQ.

MY DEAR FRIEND, April 10, 1799.

THE engagements resulting from the establishment of the Pneumatic Institution, and from a course of experiments, to which I have been obliged to pay great attention, have prevented me from acknowledging to you my obligations for the very great pleasure I received from your last excellent letter.

In experiments on Light and Heat, we have to deal with agents whose changes we are unable directly to estimate. The most we can hope for is such an arrangement of facts as will account for most of the phenomena.

The supposition of active powers common to all matter, from the different modifications of which all the phenomena of its changes result, appears to me more reasonable than the assumption of certain imaginary fluids alone endowed with active powers, and bearing the same relation to common matter, as the vulgar philosophy supposes spirit to bear to matter.

That the particles of bodies must move, or separate from each other, when they become expanded, is certain. A repulsive motion of the particles is directly the cause of expansion; and when bodies are expanded by friction, under circumstances in which there could be no heat communicated by bodies in contact, no oxidation and no diminution of capacity, I see no difficulty in conceiving the repulsive motion generated by the mechanical motion.

Your excellent and truly philosophic observations will induce me to pay greater attention to all my positions. It is only by forming theories, and then comparing them with facts, that we can hope to discover the true system of nature. I will endeavour very soon to give an answer to the remaining part of your excellent letter.

I have now just room to give you an account of

the experiments I have lately been engaged in, though they are not much connected with light and heat.

First.—One of Mr. William Coate's children accidentally discovered that two bonnet-canes rubbed together produced a faint light. The novelty of this phenomenon induced me to examine it, and I found that the canes on collision produced sparks of light, as brilliant as those from the flint and steel.

Secondly.—On examining the epidermis, I found, when it was taken off, that the canes no longer gave light on collision.

Thirdly.—The epidermis, subjected to chemical analysis, had all the properties of silex.

Fourthly.—The similar appearance of the epidermis of reeds, corn, and grasses, induced me to suppose that they likewise contained silex. By burning them carefully, and analysing their ashes, I found that they contained it in rather larger proportions than the canes.

Fifthly.—The corn and grasses contain sufficient potash to form glass with their flint. A very pretty experiment may be made on these plants with the blow-pipe. If you take a straw of wheat, barley, or hay,* and burn it, beginning at the top, and heating the ashes with the blue flame, you will obtain a perfect globule of hard glass fit for microscopic experiments.

I made a discovery yesterday which proves how necessary it is to repeat experiments. The gaseous oxide of azote is perfectly respirable when pure. It

* It is very common, after the burning of a hay-stack, to find glass in the ashes. P.

is never deleterious but when it contains nitrous gas. I have found a mode of obtaining it pure, and I breathed to-day, in the presence of Dr. Beddoes and some others, sixteen quarts of it for near seven minutes. It appears to support life longer than even oxygen gas, and absolutely intoxicated me. Pure oxygen gas produced no alteration in my pulse, nor any other material effect; whereas this gas raised my pulse upwards of twenty strokes, made me dance about the laboratory as a madman, and has kept my spirits in a glow ever since. Is not this a proof of the truth of my theory of respiration? for this gas contains more light in proportion to its oxygen than any other, and I hope will prove a most valuable medicine.

We have upwards of eighty out-patients in the Pneumatic Institution, and are going on wonderfully well.

I shall hope for the favour of a letter from you, and in my answer to it will fully inform you of our proceedings. I have just room to add that I am

Yours, with affection and respect,

HUMPHRY DAVY.

I cannot suffer the experiments with the bonnet-canes to pass, without endeavouring to infuse into the reader a portion of that admiration which I feel in relating them. They furnish a beautiful illustration of that combination of observation, experiment, and analogy, first recommended by Lord Bacon, and so strictly adopted by Davy in all his future grand researches.

In alluding to this discovery—that siliceous earth

exists generally in the epidermis of hollow plants—Davy observes in his agricultural lectures, that "the siliceous epidermis serves as a support, protects the bark from the action of insects, and seems to perform a part in the economy of these feeble vegetable tribes, similar to that performed in the animal kingdom by the shell of the crustaceous insects."

The circumstance that first led him to the investigation of the nature of *nitrous oxide*, or the *gaseous oxide of azote*, alluded to in the foregoing letter, has been thus recorded by himself. "A short time after I began the study of Chemistry, in March 1798, my attention was directed to the *dephlogisticated nitrous gas* of Priestley (nitrous oxide) by Dr. Mitchell's theory of Contagion, by which he attempted to prove that *dephlogisticated nitrous gas!* which he calls *oxide of septon*, was the principle of contagion, and capable of producing the most terrible effects, when respired by animals in the minutest quantities, or even when applied to the skin, or muscular fibre.

"The fallacy of this theory was soon demonstrated by a few coarse experiments, made on small quantities of this gas procured, in the first instance, from zinc and diluted nitrous acid. Wounds were exposed to its action; the bodies of animals were immersed in it without injury; and I breathed it, mingled in small quantities with common air, without any remarkable effects. An inability to procure it in sufficient quantities prevented me, at this time, from pursuing the experiments to any greater extent. I communicated an account of them to Dr. Beddoes."

His situation in the "Medical Pneumatic Institution" in 1799, imposing upon him the duty of investigating the physiological effects of such aëriform fluids as held out any promise of useful agency, he resumed the investigation; a considerable period, however, elapsed, before he succeeded in procuring *nitrous oxide* in a state of purity; he was therefore obliged to breathe it in mixture with oxygen gas, or common air; but as no just conclusion could be deduced from the action of an impure gas, he commenced an enquiry for the purpose of discovering a process by which it might be procured in an uncontaminated condition; when, after a most laborious investigation concerning its composition, properties, and combinations, enquiries which were necessarily extended to the different bodies connected with nitrous oxide, such as *nitrous gas*, *nitrous acid*, and *ammonia*, he was enabled, by a series of intermediate and comparative experiments, to reconcile apparent anomalies, and thus, by removing the greater number of those difficulties which had previously obscured this branch of science, to present to the chemical world the first satisfactory history of the COMBINATIONS OF OXYGEN AND NITROGEN.

Thus prepared, he proceeded to examine the action of nitrous oxide upon living beings, and to compare it with the effects of other gases upon man; and in this manner he completed its physiological, as he had already done its chemical history.

These interesting results were published in a distinct volume, in the year 1800, entitled, "Researches Chemical and Philosophical, chiefly concerning Nitrous Oxide, and its Respiration. By Humphry Davy, Superintendent of the Medical Institution."

It may be observed in passing, that the merits of this work could never have been inferred from the title-page, which its most sanguine admirers must admit to be as clumsy and unpromising an invitation as an author ever addressed to his scientific brethren.

Amongst Davy's letters to Mr. Gilbert, I find one written on a proof sheet of the chapter of contents of the above work, and which may not be uninteresting in this place.

TO DAVIES GIDDY, ESQ.

July 3, 1800.

THAT our feelings, as well as our actions, are rendered stronger and more vivid by habit, is probable from many facts, and from no one more so than that of procrastination. My much respected friend, two months after my return,* I had formed the resolution of writing to you; week after week this resolution was renewed and put off to a future day, with the hope that this day, by presenting something new, would enable me to make my letter more interesting. In vain! the feeling of procrastination, thus increased by association, at length became so strong as to prevent me from writing at all.†

* From his visit to London, as noticed at page 62.

† With respect to the metaphysical speculation contained in this paragraph of his letter, had he not written it in haste, we might presume he would have given a more exact expression to his

I have received your letter; it has awakened my duties, and has been doubly welcome, as being unexpected and undeserved.

Since my return to the Pneumatic Institution in December, I have been almost incessantly occupied, from January to April, in completing a series of experiments on Gases, and their application; and from April to the present time, in writing and printing an account of them.

I have written this letter on the table of contents of a work which will be published in the course of the month, and of which I shall take the earliest opportunity to send you a copy. This table of contents will give you a better idea of the nature and extent of the investigation, than I could possibly have given in a letter.

We have been repeating the Galvanic experiments with success. Nicholson, by means of a hundred pieces of silver and zinc, has procured a visible spark. Cruickshank has revived oxidated metals in solution, by means of the nascent hydrogen produced from the decomposition of water by the shock; and both he and Carlisle have absolutely resolved water into oxygen and hydrogen by means of it, making use of silver and platina wires. An immense field of investigation seems opened by

ideas. By the misapplied term "*Feeling of Procrastination,*" he doubtless meant to describe that aversion to labour which becomes habit by indulgence, and the perception of which, so far from increasing in vividness, actually languishes to obtuseness. To borrow an expression from Dr. Johnson, Davy, in his metaphysical speculations, not unfrequently trod upon the brink of meaning, where light and darkness begin to mingle.

this discovery: may it be pursued so as to acquaint us with some of the laws of life!

You have, undoubtedly, heard of Herschel's discovery concerning the production of heat by invisible rays emitted from the sun. By placing one thermometer within the red rays, separated by a prism, and another beyond them, he found the temperature of the outside thermometer raised more than that of the inside one.

When I first heard of Mr. Tennant's discovery,* I was very much struck by an observation which you long ago made to me, on the fertility of the Cornish lands, in which there was decomposed, *felt-spar* or *serpentine.*

Mr. Tennant spent a day here some time ago, when I mentioned your observation to him, but he could not give any solution of the phenomenon. *Quere.*—As lime and magnesia are probably both subservient to vegetation, only from supplying plants with carbonic acid, may not lime, when mingled with magnesia, in the process of vegetation, render it partially caustic, and thus enable it to destroy them?

Your observation on the scale of numbers, and the fact relative to it, are highly interesting. Reasoning on this subject would literally form the logic of generalization, or the application of one term to signify many terms, or many ideas, on which science ultimately depends. *Quere.*—How far have the first attempts at generalization arisen from acci-

* Davy here alludes to the fact of magnesian earth being prejudicial to vegetation.

dent, and how far from the resemblance between ideas?

Dr. Beddoes has always ridiculed the "*Tractors*," in common with all other reasonable men. He is about to publish a new work on the Nitrous Acid.

J. Wedgwood is returned, very little altered for the better. Coleridge is gone to reside in Cumberland; he was here the week before last, and spent much time with me, and often spoke of you with the greatest interest. Clayfield is at this moment chiefly engaged in commercial speculations. He has found a new mode of making soda, which there is every reason to believe will turn out profitable.

I hope some time in the autumn to see you, and to enjoy the well remembered pleasure of your conversation; in the mean while, I remain, with respects to your family,

Yours with sincere affection,

HUMPHRY DAVY.

In estimating the early genius of Davy, and his character as a philosopher, the style and matter of his "RESEARCHES" will afford us much assistance. The close philosophical reasoning,— the patient and penetrating industry,— the candid submission to every intimation of experiment, and the accuracy of manipulation, so remarkably displayed throughout this work, have been rarely equalled, and perhaps never surpassed.

There is scarcely to be found a more striking illustration of chemical genius, than that afforded by his chapter on the "Absorption of *Nitrous Gas by solutions of green Sulphate of Iron*."

The address with which he gradually disentangles the subject of its difficulties, and catches at every opening to truth, affords a study which may be safely recommended to the attention of every young experimentalist, as being no less instructive than it is beautiful.

The phenomena attending the absorption of *nitrous gas* by solutions of *sulphate of iron* had been examined by Vauquelin and by Berthollet, but the conclusions of these chemical philosophers were fatally infected by errors, arising from the neglected action of the atmosphere. Davy, by conducting his experiments over mercury, proved that, in the absence of air, the absorption was simply owing to a combination between the gas and the fluid; but that, on admitting air, the nitrous *gas* became nitrous *acid*, a portion of which, together with a part of the water, subsequently underwent decomposition, and gave origin to *ammonia*, and ultimately to *nitrate of ammonia*, while the iron passed into the state of a *peroxide*.

We have also to admire in this work an ardour for investigation, which even the most imminent personal danger could not repress. He may truly be said to have sought the bubble reputation in the very jaws of Death. What shall we say of that spirit which led him to inspire nitrous gas, at the hazard of filling his lungs with the vapour of *aqua fortis!* or what, of that intrepid coolness which enabled him to breathe a deadly gas, and to watch the advances of its chilling power in the ebbing pulsations at the wrist!

These experiments, however, are far too interesting and important to be related in any other than the author's own words; but it is first necessary that his trials with the *nitrous oxide* should be considered.

He found that this gas might be most conveniently, as well as most economically, prepared by the decomposition of a salt known by the name of *nitrate of ammonia*, by the application of a regulated heat; but, as the researches by which he arrived at this conclusion are recorded at length in his work, and as the most important of them are now embodied in every elementary system of chemistry, it would not only be tedious but useless, to enter into a detail of them upon this occasion.

"In April," he says, "I obtained nitrous oxide in a state of purity, and ascertained many of its chemical properties. Reflections upon these properties, and upon former trials, made me resolve to inspire it in its pure form, for I saw no other way in which its respirability, or powers, could be determined.

"I was aware of the danger of the experiment. It certainly would never have been made, if the hypothesis of Dr. Mitchell had in the least influenced my mind. I thought that the effects might, possibly, be depressing and painful; but there were many reasons which induced me to believe, that a single inspiration of a gas, apparently possessing no immediate action on the irritable fibre, could neither destroy, nor materially injure, the powers of life.

"On April 11th, I made the first inspiration of pure nitrous oxide. It passed through the bronchiæ without stimulating the glottis, and produced no uneasy sensations in the lungs.

"The result of this experiment proved that the gas was respirable, and induced me to believe that a farther trial of its effects might be made without danger.

"On April 16th, Dr. Kinglake being accidentally present, I breathed three quarts of nitrous oxide from and into a silk bag, for more than half a minute, without previously closing my nose, or exhausting my lungs. The first inspirations occasioned a slight degree of giddiness, which was succeeded by an uncommon sense of fulness in the head, accompanied with loss of distinct sensation and voluntary power,—a feeling analogous to that produced in the first stage of intoxication; but unattended by pleasurable sensation. Dr Kinglake, who felt my pulse, informed me that it was rendered quicker and fuller.

"This trial did not satisfy me with regard to its powers: comparing it with the former ones, I was unable to determine whether the operation was stimulant or depressing.

"I communicated the result to Dr. Beddoes, and on April the 17th, he was present when the following experiment was made.

"Having previously closed my nostrils, and exhausted my lungs, I breathed four quarts of the gas from and into a silk bag. The first feelings were similar to those produced in the last experiment; but in less than half a minute, the respiration being continued, they diminished gradually, and were succeeded by a sensation analogous to gentle pressure on all the muscles, attended by an highly pleasurable thrilling, particularly in the chest and in the extremities. The objects around me became dazzling, and my hearing more acute. Towards the last inspirations, the thrilling increased, the sense of muscular power became greater, and, at last, an irresistible propensity to action was indulged in: I recollect but indistinctly what followed; I know that my motions were various and violent.

"These effects very soon ceased after the respiration of the gas. In ten minutes I had recovered my natural state of mind. The thrilling in the extremities continued longer than the other sensations.

"This experiment was made in the morning; no languor or exhaustion was consequent; my feelings throughout the day were as usual, and I passed the night in undisturbed repose.

"The next morning the recollection of the effects of the gas was very indistinct; and had not remarks written immediately after the experiment recalled them to my mind, I should even have questioned their reality."

Our philosopher very naturally doubted whether some of these strong emotions might not, after all, be attributed to the enthusiasm necessarily connected with the perception of agreeable feelings, when he was prepared to expect painful sensations; but he says, that subsequent experiments convinced him that the effects were solely owing to the specific operation of the gas. He found that he could breathe nine quarts of nitrous oxide for three minutes, and twelve quarts for rather more than four; but that he could never breathe it, in any quantity, so long as five minutes. Whenever its operation was carried to the highest extent, the pleasurable thrilling, at its height about the middle of the experiment, gradually diminished, the sense of pressure on the muscles was lost, impressions ceased to be perceived, vivid ideas passed rapidly through the mind, and voluntary power was altogether destroyed, so that the mouth-piece generally dropped from his unclosed lips. When he breathed from six to seven quarts, muscular motions were produced to a great extent: sometimes he manifested his pleasure by stamping, or laughing only; at other times, by dancing round the room, and vociferating.

During the progress of these experiments, it occurred to him that, supposing *nitrous oxide* to be analogous in its operation to common stimulants, the debility occasioned by intoxication from fermented liquors ought to be increased after excitement from this gas, in the same manner as the debility produced by two bottles of wine is increased by a third. To ascertain whether this was the case, he drank a bottle of wine, in large draughts, in less than eight minutes. His usual drink, he tells us, was water; he had been little accustomed to take spirits or wine, and had never been intoxicated but once before in the course of his life. Under such circumstances, we may readily account for the powerful effects produced by this quantity of wine, and which he describes in the following manner:—

"Whilst I was drinking, I perceived a sense of fulness in the head, and throbbing of the arteries, not unlike that produced in the first stage of nitrous oxide excitement: after I had finished the bottle, this fulness increased, the objects around me became dazzling, the power of distinct articulation was lost, and I was unable to stand steadily. At this moment, the sensations were rather pleasurable than otherwise; the sense of fulness in the head, however, soon increased, so as to become painful, and in less than an hour I sunk into a state of insensibility. In this situation I must have remained for two hours, or two hours and a half. I was awakened by head-ache and painful nausea. My bodily and mental debility were excessive, and the pulse feeble and quick.

"In this state, I breathed for near a minute and a half five quarts of gas, which was brought to me by the operator for nitrous oxide; but as it produced no sensations whatever, and apparently rather increased my debility, I am almost convinced that it was, from some accident, either common air, or very impure nitrous oxide.

"Immediately after this trial, I respired twelve quarts of oxygen for nearly four minutes. It produced no alteration in my sensations at the time, but immediately afterwards I imagined that I was a little exhilarated.

"The head-ache and debility still, however, continuing with violence, I examined some nitrous oxide which had been prepared in the morning, and finding it very pure, I respired seven quarts of it for two minutes and a half. I was unconscious of head-ache after the third inspiration; the usual pleasurable thrilling was produced, voluntary power was destroyed, and vivid ideas rapidly passed through my mind; I made strides across the room, and continued for some minutes much exhilarated; but languor and depression, not very different in degree from those existing before the experiment, succeeded; they however gradually went off before bed-time.

"This experiment proved, that debility from intoxication was not increased by excitement from nitrous oxide. The head-ache and depression would probably have continued longer, had it not been administered."

The same work contains an account of many other trials; but sufficient has been extracted to show the zeal and intrepidity with which he conducted his researches. To withhold, however, the testimony which several other scientific persons have given, with respect to the intoxicating influence of this gas, would be to deprive the reader of some very amusing descriptions.

First appears Mr. W. Tobin, who tells us that he soon found his nervous system agitated by the highest sensations of pleasure, but which were difficult of description. When the bags were exhausted and taken from him, he suddenly started from his chair, and vociferating with pleasure, made towards those that were present, as he wished they should participate in his feelings. He struck gently at Davy, and a stranger entering the room at the same moment, he made towards him, and gave him several blows, but he adds, it was more in the spirit of good-humour, than in that of anger. He then ran through different rooms in the house, and at last returned to the laboratory, somewhat more composed, although his spirits continued much elevated for some hours after the experiment; he felt, however, no consequent depression, either in the evening or day following. Upon another occasion, he states that his sensations were superior to any thing he ever before experienced; his step was firm, and all his muscular power increased. His nerves were more alive to every surrounding impression; he threw himself into several theatrical attitudes, and traversed the laboratory with a quick step, while his mind was elevated to a most sublime height: he says that "it is giving but a faint idea of his feelings to say, that they resembled those produced by a representation of an heroic scene on the stage, or by reading a sublime passage in poetry, when circumstances contribute to awaken the finest sympathies of the soul." The influence, however, of this inspiring agent appears to have been as transitory as its effects were vivid; for he afterwards observes, "I have seldom lately experienced vivid sensations. The pleasure produced by the gas is slight and tranquil, and I rarely feel sublime emotions, or increased muscular power."

The first time that Mr. Clayfield breathed the gas, it produced feelings analogous to those of intoxication. He was for some time unconscious of existence, but at no period of the experiment were his sensations agreeable; a momentary nausea followed, but unconnected with languor or head-ache.

In a subsequent trial, it would appear that he did experience certain thrillings which were highly pleasurable.

The account given by Dr. Kinglake agrees pretty much with those already cited. He adds, however, that the inspiration of the gas had the further effect of reviving rheumatic irritations in the shoulder and knee-joints, which had not been previously felt for many months.

Next appears Mr. Southey, the Laureate. The reader will no doubt be prepared to hear that the nitrous oxide transported him, at least, to the summit of Parnassus ;—by no means : he laughed when the bag was removed from his mouth, but it may be fairly questioned whether this might not have been an expression of joy at the terrors he had escaped ; for he freely confesses that he could not distinguish between the first feelings it occasioned, and an apprehension of which he was unable to divest himself.

The first time Mr. Coleridge inspired the nitrous oxide, he felt a highly pleasurable sensation of warmth over his whole frame : he adds, that the only motion which he felt inclined to make, was that of laughing at those who were looking at him : a symptom as equivocal, perhaps, as that exhibited by the Laureate.

A number of other accounts are given, but those already related are perhaps sufficient to establish the fact, that the gas in question possesses an intoxicating quality, to which the enthusiasm of persons submitting to its operation has imparted a character of extravagance wholly inconsistent with truth.

It will be admitted that there must have been something singularly ludicrous in the whole exhibition. Imagine a party of grave philosophers, with bags of silk tied to their mouths, stamping, roaring, and laughing about the apartment; it is scarcely possible to conceive a richer subject for the pencil of a Bunbury. We cannot then be surprised at any terms of ridicule in which a stranger, witnessing such an operation, might describe it. M. T. Fievée* appears to have considered the practice as a national vice, and whimsically introduces it amongst the catalogue of follies to which he considers the English nation to be addicted.

Taking leave of these laughing philosophers, we must now proceed to a much more serious branch of the subject of Pneumatic Medicine. "Having observed," says Davy, "that no painful effects were produced by the application of nitrous gas to the bare muscular fibre, I began to imagine that this gas might also be breathed with impunity, provided it were possible in any way to free the lungs of common air before inspiration, so as to prevent the formation of nitrous acid.

"On this supposition, during a fit of enthusiasm produced by the respiration of nitrous oxide, I resolved to endeavour to breathe nitrous gas: one hundred and fourteen cubic inches of it were accordingly introduced into the large mercurial air-holder; two small silk bags of the capacity of seven quarts were filled with nitrous oxide.

"After a forced exhaustion of my lungs, my nose being accurately closed, I made three inspirations and expirations of nitrous oxide in one of the bags, in order to free my lungs, as much as possible, from atmospheric oxygen; then, after a full expiration of the nitrous oxide, I transferred my lips from the mouthpiece of the bag to that of the air-holder, and, turning the stopcock, attempted to inspire the nitrous gas. In passing through my mouth and

* Lettres sur l'Angleterre, 1802.

fauces, it tasted astringent and highly disagreeable; it occasioned a sense of burning in the throat, and produced a spasm of the epiglottis, so painful as to oblige me to desist immediately from attempts to inspire it. After removing my lips from the mouth-piece, when I opened them to inspire common air, *nitrous acid* was immediately formed in my mouth, which burnt the tongue and palate, injured the teeth, and produced an inflammation of the mucous membrane, which lasted for some hours.

"As, after the respiration of nitrous oxide, a small portion of the residual atmospheric air always remained in the lungs mingled with the gas, so is it probable that, in the experiment just related, a minute portion of nitrous acid was formed; and, if so, I perhaps owe the preservation of my life to the circumstance; for, supposing that I had succeeded in taking a full inspiration of nitrous gas, and even that it had not produced any positive effects, it is not likely that I should, by breathing nitrous oxide, have so completely freed my lungs from it, as to have prevented the formation of nitrous acid, when I again inspired common air. I never design again to attempt so rash an experiment."

His attempt to breathe carburetted hydrogen gas was scarcely less terrific and appalling.

"Mr. Watt's observations on the respiration of diluted *hydro-carbonate* by man, and the experiments of Dr. Beddoes on the destruction of animals by the same gas, proved that its effects were highly deleterious.

"As it destroyed life, apparently by rendering the muscular fibre inirritable, without producing any previous excitement, I was anxious to compare its sensible effects with those of nitrous oxide, which at this time I believed to destroy life by producing the highest possible excitement.

"In the first experiment, I breathed for nearly a minute three quarts of *hydro-carbonate*, mingled with nearly two quarts of atmospheric air.* It produced a slight giddiness, pain in the head, and a momentary loss of voluntary power; my pulse was rendered much quicker and more feeble. These effects, however, went off in five minutes, and I had no return of giddiness.

"Emboldened by this trial, I introduced into a silk bag four quarts of gas nearly pure, which was carefully produced from the decomposition of water by charcoal an hour before, and which had a very strong and disagreeable smell.

"My friend Mr. James Tobin, junior, being present, after a forced exhaustion of my lungs, the nose being accurately closed, I made three inspirations and expirations of the hydro-carbonate. The first inspiration produced a sort of numbness and loss of feeling in the chest, and about the pectoral muscles. After the second, I lost all power of perceiving external things, and had no distinct sensation, except that of a terrible oppression on the chest. During the third expiration, this feeling subsided, I seemed sinking into annihilation, and had just power enough to cast off the mouthpiece from my unclosed lips.

* "I believe it had never been breathed before by any individual in a state so little diluted."

"A short interval must have passed, during which I respired common air, before the objects around me were distinguishable. On recollecting myself, I faintly articulated, '*I do not think I shall die.*' Placing my finger on the wrist, I found my pulse thread-like, and beating with excessive quickness. In less than a minute, I was able to walk, and the painful oppression on the chest directed me to the open air.

"After making a few steps, which carried me to the garden, my head became giddy, my knees trembled, and I had just sufficient voluntary power to throw myself on the grass. Here the painful feelings of the chest increased with such violence as to threaten suffocation. At this moment I asked for some nitrous oxide. Mr. Dwyer brought me a mixture of that gas with oxygen, and I breathed it for a minute, and believed myself recovered.

"In five minutes the painful feelings began gradually to diminish; in an hour they had nearly disappeared, and I felt only excessive weakness, and a slight swimming of the head. My voice was very feeble and indistinct.

"I afterwards walked slowly for half an hour with Mr. Tobin, and on my return was so much stronger and better as to believe that the effects of the gas had entirely passed off; though my pulse was 120, and very feeble. I continued without pain for nearly three quarters of an hour, when the giddiness returned with such violence as to oblige me to lie on the bed; it was accompanied with nausea, loss of memory, and deficient sensation.

"In about an hour and a half, the giddiness went

off, and was succeeded by an excruciating pain in the forehead, and between the eyes, with transient pains in the chest and extremities.

"Towards night these affections gradually diminished; and at ten no disagreeable feeling, except weakness, remained. I slept sound, and awoke in the morning very feeble and very hungry. No recurrence of the symptoms took place, and I had nearly recovered my strength by the evening.

"I have been minute in the account of this experiment, because it proves, that *hydro-carbonate* acts as a sedative; that is, it produces diminution of vital action, and consequent debility, without previously exciting. There is every reason to believe that, had I taken four or five inspirations, instead of three, they would have destroyed life immediately, without producing any painful sensation."

The scientific and medical world are alike indebted to Davy for this daring experiment; and, if the precautions it suggests be properly attended to, it may become the means of preserving human life. The experiment is also valuable as affording support to physiological views, with which its author was probably not acquainted.

In the first place, it may be necessary to apprize some of my readers, that the "*hydro-carbonate*" here spoken of, differs very little from the gas now so generally used to illuminate our streets and houses. We have just seen how deadly are its qualities, and that even in a state of extreme dilution it will affect our sensations. The question then necessarily suggests itself, how far this gas can be sarely introduced into the interior of our apartments?

Did we not possess any direct evidence upon the subject, the answer would be sufficiently obvious, since it is impossible so to conduct its combustion, that a portion shall not escape unburnt. Such is the theory; but what is our experience upon the subject?—that pains in the head, nausea, and distressing languor, have been repeatedly experienced in our theatres and saloons, by persons inhaling the unburnt gas; that the atmosphere of a room, although spacious and empty, will, if lighted with gas, convey a sense of oppression to our organs of respiration, as if we were inhaling an air contaminated with the breath of a hundred persons.

In the next place, Davy's experiment is important, inasmuch as it proves that, in cases of asphyxia, or suspended animation, there exists a period of danger after the respiration has been restored, and the circulation re-established, at which death may take place, when we are the least prepared to expect it.

Bichat has shown that, when dark-coloured blood is injected into the vessels of the brain by means of a syringe connected with the carotid artery, the functions of the brain become immediately disturbed, and in a short time entirely cease: the effect is precisely similar, whether the dark-coloured blood be transmitted to the brain by the syringe of the experimentalist, or by the heart itself. Thus in cases of asphyxia, the dark-coloured blood which has been propelled through the vessels during the suspension or imperfect performance of respiration, acts like a narcotic poison upon the brain; and no sooner, therefore, does it extend its malign influence to that organ, than deleterious effects are produced, and the animal, after apparent recovery, falls into a state of stupor, the pupils of the eyes become dilated, the respiration laborious, the muscles of the body convulsed, and it speedily dies, *poisoned by its own blood.*

We are much indebted to Mr. Brodie for a series of experiments in confirmation of these views; and a very interesting case occurred some time since, in the neighbourhood of Windsor, which is well calculated for their illustration. A corporal in the Guards, whose name, if I am not mistaken, was Schofield, was seized with cramp as he was bathing in the Thames, and remained for several minutes under water. By judicious assistance, however, he was recovered, and appeared to those about him to be free from any danger, when he was attacked by convulsions and expired. Had the respiration been artificially supported at this period, so as to have maintained the action of the heart until the black blood had returned from the brain, the life of the soldier might possibly have been saved.

In the experiment which has given origin to these reflections, Davy distinctly states, that after having recovered from the *primary* effects of the carburetted hydrogen gas, and taken a walk with his friend Mr. Tobin, he was again seized with violent giddiness, attended with nausea and loss of sensation. The imperfectly oxygenized or dark-coloured blood had evidently affected the brain, and his life, at this period, was probably in greater jeopardy than in any other stage of the experiment.

Nothing daunted by the dangers to which the

preceding experiments had exposed him, Davy did not allow more than a week to elapse before he attempted to respire *fixed air*, or *carbonic acid gas*, but it was in vain that he made voluntary efforts to draw it into the windpipe; for, the moment the epiglottis was raised a little, such a painful irritation was induced as instantly to close it spasmodically on the glottis; and thus, in repeated trials, was he prevented from taking a single particle of carbonic acid into the lungs. When, however, the gas was diluted with a little more than double its volume of common air, he was enabled to breathe it for nearly a minute, when it produced a slight degree of giddiness, and an inclination to sleep.*

It may perhaps appear extraordinary to the reader of the "RESEARCHES," that although they were published not more than eighteen months after the appearance of his "Essays on Heat and Light," no allusion is made in them either to his theory or to his new nomenclature. In relating his experiments upon Respiration, he employs the conventional lan-

* It would thus appear that carbonic acid, in its most concentrated form, may kill by exciting a spasmodic action, in which the epiglottis is closed, and the entrance of air into the lungs altogether prevented. In a diluted form, it may destroy by its specific influence upon the blood, which would seem to be of a highly sedative character. Death produced by such an agent is probably attended with little or no suffering. The younger Berthollet destroyed his life by inclosing himself in an atmosphere of this description; and on commencing his fatal experiment, he registered all the successive feelings he experienced, which were such as would have been occasioned by a narcotic:—a pause, and then an almost illegible word occurred: it is presumed that the pen dropped from his hand,—and he was no more.

guage of the schools, and the word "*phos-oxygen*" does not once occur in the volume. This is fully explained in a communication made by him to Mr. Nicholson, and which was printed in his Journal a short time after the publication of his Essays in the West Country Contributions; in which he says,— "As facts have occurred to me with regard to the decomposition of bodies, which I had supposed to contain light, without any luminous appearance, I beg to be considered as a *sceptic* with respect *to my own* particular theory of the combinations of light, until I shall have satisfactorily explained those anomalies by fresh experiments. On account of this scepticism, and for other reasons, I shall in future use the common nomenclature; excepting that, as my discoveries concerning the gaseous oxide would render it highly improper to call a principle, which in one of its combinations is capable of being absorbed by venous blood, and of increasing the powers of life, *azote*,—I shall name it, with Dr. Pearson, Chaptal, and others, NITROGENE; and the *gaseous oxide of azote* I shall call NITROUS OXIDE."

The same feeling is expressed at the conclusion of his Third Research.—"It would be easy to form theories referring the action of blood impregnated with *nitrous oxide*, to its power of supplying the nervous and muscular fibre with such proportions of condensed nitrogen, oxygen, light, or ethereal fluid, as enabled them more rapidly to pass through those changes which constitute their life; but we are unacquainted with the composition of dead organized matter; and new instruments of experiment, and new modes of research, must be found, before we can ascertain even our capabilities of discovering the laws of life."

There is one circumstance connected with the views entertained in this work which must not be passed over without notice. In several passages he advocates the theory of the atmosphere being a *chemical compound* of oxygen and nitrogen; whereas, in later years, he was amongst the first to insist upon its being simply a mechanical mixture of these gases.

In consequence of the highly deleterious experiments which have been already described, and of the constant labours of the laboratory, and the repeated inhalation of acid and other vapours, his health began visibly to decline, and he retired into Cornwall, where he informs us that "the associations of ideas and feelings, common exercise, a pure atmosphere, luxurious diet, and a moderate indulgence in wine, in the course of a month, restored him to health and vigour."

I find an allusion to this visit in a letter from his sister. "He had," she says, "written to his mother of his intention to visit her, but before the post had quitted Bristol, he was already on his way to Penzance, and would have reached it before his letter, had not his aunt, on whom he called in the neighbouring town of Marazion, struck with his appearance of ill health, insisted on his remaining there till the next day, lest his mother should be doubly alarmed at his unexpected visit and altered looks." Miss Davy adds, "This one fact will serve, at the same time, to illustrate his attachment to home, and the impetuosity of his mind, which never rested till the object he proposed was accomplished."

The following letter is inserted in this place, for the purpose of fixing the period at which he first ascertained those new facts in Voltaic electricity, which formed the basis of a future communication to the Royal Society, and which may be said to have paved the way to his grand discoveries in that branch of science;—the dawning of that glorious day, which we shall presently view in all its splendour and glory.

There is, moreover, something extremely interesting in receiving from himself a simple and unadorned statement of results, as they successively presented themselves to his observation — "Truths plucked as they are growing, and delivered to you before their dew is brushed off."

TO DAVIES GIDDY, ESQ.

Pneumatic Institution, October 20, 1800.

Be assured, my respected friend, that your last letter, though short, was highly gratifying to me. At the moment it was brought to me, I was about to depart with King and Danvers on an excursion to the banks of the Wye. Our design was to see Tintern Abbey by moonlight, and it was perfectly accomplished.

After viewing for three hours all the varieties of light and shade which a bright full moon and a blue sky could exhibit in this beautiful ruin, and after wandering for three days among the many-coloured woods and rocks surrounding the river between Monmouth and Chepstow, we arrived on the fourth day at Bristol, having to balance against the pleasure of the tour, the fatigue of a stormy voyage down the Wye, across the mouth of the Severn, and up the Avon.

On analysing, after our return, specimens of the air collected from Monmouth, from the woods on the banks of the Wye, and from the mouth of the Severn, there was no perceptible difference; they were all of similar composition to the air in the middle of Bristol; that is, they contained about twenty-two per cent. of oxygen. The air from the bladders of some sea-weed, apparently just cast on shore, at the Old Passage, likewise gave the same results; so that, comparing these experiments with those made by Cavendish, Berthollet, &c. and by myself on other occasions, at different temperatures, in different weather, and with different winds, I am almost convinced that the whole of the lower stratum of the atmosphere is of uniform composition.

No test can be more fallacious and imperfect than nitrous gas, on account of the different composition of nitrous acid, formed in the different manipulations of eudiometrical experiments.

The eudiometer that I have lately employed gives, in a few minutes, the proportions of oxygen without correction.

In pursuing experiments on galvanism, during the last two months, I have met with unexpected and unhoped-for success. Some of the new facts on this subject promise to afford instruments capable of destroying the mysterious veil which Nature has thrown over the operations and properties of ethereal fluids.

Galvanism I have found, by numerous experiments, to be *a process purely chemical*, and to depend wholly on the oxidation of metallic surfaces, having different degrees of electric conducting power.

Zinc is incapable of decomposing *pure* water; and if the zinc plates be kept moist with *pure* water, the galvanic pile does not act; but zinc is capable of oxidating itself when placed in contact with water, holding in solution either oxygen, atmospheric air, or nitrous or muriatic acid, &c.: and under such circumstances, the galvanic phenomena are produced, and their intensity is in proportion to the rapidity with which the zinc is oxidated.

The galvanic pile only acts for a few minutes, when introduced into hydrogen, nitrogen, or hydrocarbonate; that is, only as long as the water between its plates holds some oxygen in solution: immerse it for a few moments in water containing air, and it acts again.

It acts very vividly in oxygen gas, and less so in the atmosphere. When its plates are moistened by marine acid, its action is very powerful, but infinitely more so when nitrous acid is employed. Five plates with nitrous acid gave sparks equal to those of the common pile. From twenty plates the shock was insupportable.

I had almost forgotten to mention, that charcoal is a good galvanic exciter, and decomposes water, like the metals, in the pile; but I must stop, without being able to expatiate on the connection which is now obvious between galvanism and some of the phenomena of organic motion. I never consider the subject without having forcibly impressed upon my imagination your observations* on the science of the ethereal fluids, and I cannot help flattering myself that this age will see your predictions verified. I remain with sincere respect and affection,

Yours,

HUMPHRY DAVY.

That a work, of the character of the "RESEARCHES," replete with ingenious novelty, and rich in chemical discovery, proceeding from the pen of so young a man, should have excited very general admiration in the philosophic world, is a circumstance that cannot surprise us; but in a majority of cases, precocious merit enjoys only an ephemeral popularity; the sensations it excites are too vivid to be permanent, and the individual sinks into an obscurity rendered ten times more profound by the brilliancy of the flash which preceded it; but every event of Davy's life would appear as if created, and directed for his welfare, by some presiding genius, whose activity in throwing opportunities in his way was rivalled only by the address with which he converted them to his advantage. Fortune and talent, then, were both equally engaged in accomplishing the elevation of Davy, and it is probable that eminent success generally requires a combination of these elements for its production, and that

* On conversing with Mr. Gilbert on the above passage, I understand that it is an allusion to his opinion, that the discovery of Galvanic power would ultimately lead to a knowledge of the nature of light and heat.

the maxim of Plautus is therefore as remote from truth as that of Theophrastus, the one assigning all to fortune, the other all to talent.

The experiments to which allusions have been frequently made during the present chapter, favourably as they were received, might have shared the fate of many other discoveries which did not admit of an immediate and obvious application to the purposes of common life; for statistical value is a necessary passport to popular favour. Fortunately, however, for Davy, before the vivid impression produced by his new work had lost the glow of novelty, Count Rumford was anxiously seeking for some rising philosopher, who might contribute his energies towards the support, and farther increase, of the chemical fame of the recently established "INSTITUTION OF GREAT BRITAIN."

It is not surprising that his attention should have been readily directed to one whose genius had been so lately displayed, and whose views regarding Caloric * were in such exact conformity with his own opinions.

* Mr. Gilbert no sooner discovered the tendency of Davy's opinions respecting the immateriality of Caloric, than he urged him to communicate them to Count Rumford, but he considered himself pledged to Dr. Beddoes, and his Essays were accordingly printed in the West Country Contributions. Count Rumford, it may be observed, maintained that Caloric, like *Phlogiston*, was merely a creature of the chemist's imagination, and had no real existence. He considered heat as nothing more than the motions of the constituent particles of bodies amongst themselves,—an hypothesis which has no claims to novelty, being perhaps one of the most ancient on record.—See his paper on Heat, *Phil. Trans.* for 1804.

As the philosophical public must feel a lively interest in every incident connected with a transaction so important to the interests of science, as that by which Davy was placed in the chemical chair of the Institution, I am fortunate in being able, through the kindness of his two friends, Mr. Thomson and Mr. Underwood, to present a clear and satisfactory statement of all its circumstances and details.

CHAPTER III.

Count Rumford negotiates with Mr. Underwood on the subject of Davy's appointment to the Royal Institution.—Terms of his engagement communicated in a letter to Mr. Gilbert.—Davy arrives, and takes possession of his apartments.—He receives various mortifications.—He is elected a member of the Tepidarian Society.—Is appointed Lecturer instead of assistant.—He makes a tour in Cornwall with Mr. Underwood.—Anecdotes.—His Poem on Spinosism.—His letter to Mr. Gilbert, communicating a galvanic discovery.—He commences his first grand course of lectures. — His brilliant success. — A letter from Mr. Purkis.—Davy's style criticised.—His extraordinary method of experimenting.—Davy and Wollaston compared as experimentalists.—The style of Davy as a lecturer and a writer contrasted

It may be readily supposed that the prominent situation held by Davy at Bristol, as well as the merited celebrity of his writings, must have rendered his name familiar to all the leading philosophers of the day. It were vain, therefore, to enquire through what channel the echo of his fame first reached the ear of Count Rumford;* it is suffi-

Amongst other celebrated chemists who had become acquainted with Davy at Bristol, and subsequently spoken of his extraordinary genius, was Dr. Hope. He informs me that Count Rumford had applied to him to find some chemist who would undertake the office of lecturer at the Institution; but that he failed

cient to state that Mr. Underwood, a gentleman ardently attached to science, and devoted to the interests of the Royal Institution, was amongst the first to urge the expediency of inviting him to London as a public lecturer. Mr. Underwood, in a letter lately addressed to me from Paris, says, "In consequence of several conversations with Count Rumford, on the subject of Davy's superior talents, and the advantages that would accrue to the Institution from engaging him as a lecturer, the Count called upon me on the 5th of January 1801, having received from the Managers of the Institution full powers to negotiate upon the subject. On this occasion, however, I thought it advisable to introduce the Count to Mr. James Thomson, as being the more eligible person to treat in behalf of Davy, not only on account of his greater intimacy with him, but because, not being a proprietor, he was unconnected with the interests of the Institution."

Mr. Thomson, who saw the prospect of honour and emolument thus opened for his friend, after a satisfactory interview with Count Rumford, immediately wrote to Davy, with an earnest recommendation that he should, without loss of time, come to town, and conclude an arrangement thus auspiciously commenced.

Davy, with his characteristic ardour, answered

in discovering such a person as he could, with propriety, introduce; some time afterwards, however, he became acquainted with Davy, and having soon perceived his talents, recommended him without any hesitation to the patronage of the Count. This circumstance, combined with several others, no doubt might have had its influence in deciding the fate of Davy.

the letter in person. He was introduced to the Managers, and the preliminary arrangements were soon completed; the nature of which is disclosed by himself in the following letter to Mr. Gilbert.

Hotwells, March 8, 1801.

I CANNOT think of quitting the Pneumatic Institution, without giving you intimation of it in a letter; indeed, I believe I should have done this some time ago, had not the hurry of business, and the fever of emotion produced by the prospect of novel changes in futurity, destroyed to a certain extent my powers of consistent action.

You, my dear Sir, have behaved to me with great kindness, and the little ability I possess you have very much contributed to develope; I should therefore accuse myself of ingratitude, were I to neglect to ask your approbation of the measures I have adopted with regard to the change of my situation, and the enlargement of my views in life.

In consequence of an invitation from Count Rumford, given to me with some proposals relative to the Royal Institution, I visited London in the middle of February, where, after several conferences with that gentleman, I was invited by the Managers of the Royal Institution to become the Director of their laboratory, and their Assistant Professor of chemistry; at the same time I was assured that, within the space of two or three seasons, I should be made sole Professor of Chemistry, still continuing Director of the laboratory.

The immediate emolument offered was sufficient for my wants; and the sole and uncontrolled use of the apparatus of the Institution, for private experiments, was to be granted me.

The behaviour of Count Rumford, Sir Joseph Banks, Mr. Cavendish, and the other principal managers, was liberal and polite; and they promised me any apparatus that I might need for new experiments.

The time required to be devoted to the services of the Institution was but short, being limited chiefly to the winter and spring. The emoluments to be attached to the office of sole Professor of Chemistry are great; and, above all, the situation is permanent, and held very honourable.

These motives, joined to the approbation of Dr. Beddoes, who with great liberality has absolved me from my engagements at the Pneumatic Institution, and the strong wishes of most of my friends in London and Bristol, determined my conduct.

Thus I am quickly to be transferred to London, whilst my sphere of action is considerably enlarged, and as much power as I could reasonably expect, or even wish for at my time of life, secured to me without the obligation of labouring at a profession.

The Royal Institution will, I hope, be of some utility to society. It has undoubtedly the capability of becoming a great instrument of moral and intellectual improvement. Its funds are very great. It has attached to it the feelings of a great number of people of fashion and property, and consequently may be the means of employing, to useful purposes, money which would otherwise be squandered in luxury, and in the production of unnecessary labour.

Count Rumford professes that it will be kept distinct from party politics; I sincerely wish that such may be the case, though I fear it.* As for myself, I shall become attached to it full of hope, with the resolution of employing all my feeble powers towards promoting its true interests.

So much of my paper has been given to pure egotism, that I have but little room left to say any thing concerning the state of science, and the public mind in town; unfortunately, there is little to say. I have heard of no important discoveries. In politics, nothing seems capable of exciting permanent interest. The stroke of poverty, though severely felt, has been a torpedo, benumbing all energy, and not irritating and awakening it, as might have been expected.

Here, at the Pneumatic Institution, the nitrous oxide has evidently been of use. Dr. Beddoes is proceeding in the execution of his great popular physiological work, which, if it equals the plan he holds out, ought to supersede every work of the kind.

I have been pursuing Galvanism with labour, and some success. I have been able to produce galvanic power from simple plates, by effecting on them different oxidating and de-oxidating processes; but on this point I cannot enlarge in the small remaining space of paper.

* In England, politics so constantly mix themselves up with all our institutions, while science unfortunately finds so few disciples and patrons in the ranks of aristocracy, that every new society is viewed with jealousy and party spirit. Johnson says, in his Life of Addison—"It has been suggested that the Royal Society was instituted soon after the Restoration, to divert the attention of the people from public discontent."—P.

Your remark concerning *negative* Galvanism, and de-oxidation, is curious, and will most probably hold good.

It will give me much pleasure to see your mathematical Paper* in the Philosophical Transactions, but it will be, unfortunately, to me the pleasure of *blind* sympathy, though derived from the consciousness that you ought to be acting upon, and instructing the world at large.

It will give me sincere pleasure to hear from you, when you are at leisure. After the 11th I shall be in town—my direction, Royal Institution, Albemarle Street. I am, my dear friend, with respect and affection, Yours,

HUMPHRY DAVY.

The first notice of Davy's name, in the Minute Book of the Royal Institution, occurs in the Report adopted at a Meeting of the Managers, on Monday the 16th of February 1801.

"Resolved—That Mr. Humphry Davy be engaged in the service of the Royal Institution, in the capacities of Assistant Lecturer in Chemistry, Director of the Laboratory, and Assistant Editor of the Journals of the Institution, and that he be allowed to occupy a room in the house, and be furnished with coals and candles; and that he be paid a salary of one hundred guineas per annum."

On the 16th of March 1801, after reporting that "a room had been prepared and furnished for Davy," the Minute proceeds to state that "Mr.

* He alludes to some calculations connected with light, and the imponderable fluids.

Davy had arrived at the Institution on Wednesday the 11th of March, and taken possession of his situation."

It is a curious fact, that the first impression produced on Count Rumford by Davy's personal appearance, was highly unfavourable to the young philosopher, and he expressed to Mr. Underwood his great regret at having been influenced by the ardour with which his suit had been urged; and he actually would not allow him to lecture in the Theatre, until he had given a specimen of his abilities in the smaller lecture-room. His first lecture, however, entirely removed every prejudice which had been formed; and at its conclusion the Count emphatically exclaimed—"Let him command any arrangements which the Institution can afford." He was accordingly, on the very next day, promoted to the great Theatre.

Davy's uncouth appearance and address subjected him to many other mortifications on his first arrival in London. There was a smirk on his countenance, and a pertness in his manner, which, although arising from the perfect simplicity of his mind, were considered as indicating an unbecoming confidence. Johnson, the publisher, as many of my readers will probably remember, was in the custom of giving weekly dinners to the more distinguished authors and literary stars of the day. Davy, soon after his appointment, was invited upon one of these occasions, but the host actually considered it necessary to explain, by way of apology, to his company, the motives which had induced him to introduce into their society a person of such humble pretensions. At this dinner a circumstance occurred, which must have been very mortifying to the young philosopher. Fuseli was present, and, as usual, highly energetic upon various passages of beauty in the poets, when Davy most unfortunately observed, that there were passages in Milton which he could never understand. "Very likely, very likely, Sir," replied the artist in his broad German accent, "but I am sure that is not Milton's fault."

On the 7th of April, he was elected a member of a society which consisted of twenty-five of the most violent republicans of the day; it was called the "*Tepidarian* Society," from the circumstance of nothing but tea being allowed at their meetings, which were held at Old Slaughter's Coffee House in Saint Martin's Lane. To the influence of this society, Mr. Underwood states that Davy was greatly indebted for his early popularity. Fame gathers her laurels with a slow hand, and the most brilliant talents require a certain time for producing a due impression upon the public; the *Tepidarians* exerted all their personal influence to obtain an audience before the reputation of the lecturer could have been sufficiently known to attract one.

Although the acquaintance between Davy and Count Rumford commenced so inauspiciously, they very soon became friends, and mutually entertained for each other the highest regard.

Davy's improved manners and naturally simple habits, at this period, were highly interesting and exemplary: towards his old friends he conducted himself with the greatest amenity, and frequently consulted them upon certain points connected with his new station in society. The following anecdote was communicated by Mr. Underwood.—"I introduced him," says he, "to my old friend, the excellent Sir Henry Englefield, who was the first intimate acquaintance Davy had formed in the higher circles; he was received by him with all that warmth of manner, and kindness of feeling, which so eminently distinguished him. Shortly after this introduction, Sir Harry sent him an invitation to meet me at dinner. Davy found himself unable to frame an answer to his satisfaction, and fearing he might betray his ignorance of etiquette, ran to my house, and greatly alarmed my mother by the extreme anxiety he displayed, and the manner in which he entreated her to send me to him the moment I returned. I went and found him cudgelling his brains to produce this first attempt at fashionable composition; a dozen answers were on his table, and he was in the highest degree excited and annoyed."

It would appear that, immediately after his arrival at the Royal Institution, he entered upon the duties of his station, and performed them so greatly to the satisfaction of the Managers, that, at a Meeting held on the first of June, not more than six weeks afterwards, the following Resolutions were passed.

"Resolved—That Mr. Humphry Davy, Director of the Chemical Laboratory, and Assistant Lecturer in Chemistry, has, since he has been employed at the Institution, given satisfactory proofs of his talents as a Lecturer.

"Resolved—That he be appointed, and in future denominated, Lecturer in Chemistry at the Royal

Institution, instead of continuing to occupy the place of *Assistant* Lecturer, which he has hitherto filled."

From an examination of the Minute Book, it appears that Dr. Garnett, whose health had long been on the decline, resigned his professorship on the 15th of June,* and that on the 6th of July in the same year, Dr. Young was engaged as Professor of Natural Philosophy, Editor of the Journals, and general Superintendent of the House, at the salary of 300*l.* per annum.

With Dr. Garnett he had lived on terms of great intimacy; with his successor he associated with less ease and freedom.

At a meeting of the Managers, also held in July, several Resolutions were passed to the following effect.

"Resolved—That a Course of Lectures on the Chemical Principles of the Art of Tanning be given by Mr. Davy. To commence the second of November next; and that respectable persons of the trade, who shall be recommended by Proprietors of the Institution, be admitted to these lectures gratis.

"Resolved further—That Mr. Davy have permission to absent himself during the months of July, August, and September, for the purpose of making himself more particularly acquainted with the practical part of the business of tanning, in order to prepare himself for giving the above-mentioned course of lectures."

Davy, it would appear, availed himself of the permission granted to him by the above resolution

* He delivered his farewell Lecture on the 9th of the same month.

of the Managers, and during the interval visited his native country.

He had arranged with his friend Underwood to make a tour through Cornwall; but as it was his wish to remain at Bristol for a few days, on his way to the West, it was agreed between them that they should meet at Penzance.

Davy, however, became impatient, and wrote the following letter to his friend, a composition of much wildness, and obnoxious to the suspicion of Spinosism.

MY DEAR UNDERWOOD,

THAT part of Almighty God which resides in the rocks and woods, in the blue and tranquil sea, in the clouds and sunbeams of the sky, is calling upon thee with a loud voice: religiously obey its commands, and come and worship with me on the ancient altars of Cornwall.

I shall leave Bristol on Thursday next, possibly before, so that by this day week I shall probably be at Penzance. Ten days or a fortnight after, I shall expect to see you, and to rejoice with you.

We will admire together the wonders of God,—rocks and the sea, dead hills and living hills covered with verdure. Amen.

Write to me immediately, and say when you will come. Direct H. Davy, Penzance. Farewell, Being of energy!

Your's with unfeigned affection,

H. DAVY.

Mr. Underwood transmitted to me the above letter with the following extract from his journal.

"On the 25th I went to Bristol, and on the 30th arrived at Mrs. Davy's at Penzance. On the 1st of August we set off on a pedestrian excursion, and proceeded along the edge of the cliffs, round the Land's End, Cape Cornwall, Saint Just, and Saint Ives, to Redruth, and thence back to Penzance.

"Two days afterwards we again started, and trudged along the shore to the Lizard. Kynance Cove had from the commencement of our intimacy been a daily theme of his conversation. No epithets were sufficiently forcible to express his admiration at the beauty of the spot: the enthusiastic delight with which he dwelt upon the description of the Serpentine rocks, polished by the waves, and reflecting the brightest tints from their surfaces, seemed inexhaustible, and when we had arrived at the spot he appeared absolutely entranced.

"During these excursions his conversation was most romantic and poetical. His views of Nature, and her sublime operations, were expressed without reserve, as they rapidly presented themselves to his imagination: they were the ravings of genius; but even his nonsense was that of a superior being."

At the village of Mullion, a little incident occurred, which evinced the existence of that gastronomic propensity which, in after years, displayed itself in a wider range of operations. The tourists had, on their road, purchased a fine large bass of a fisherman, with the intention of desiring the landlady to dress it. On arriving at the inn, Mr. Underwood retired into a room for the purpose of making some notes in the journal he regularly kept. Davy had disappeared. In the course of a few minutes a most tremendous uproar was heard in the kitchen, and the indignant vociferations of the hostess, which, even with all the advantages of Cornish recitative, was not of the most melodious description, became fearfully audible. Davy, it seems, had volunteered his assistance in making the sauce and stuffing for the aforesaid bass; and had he not speedily retreated, his services would have been rewarded, not according to the scientific practice of appending a string of letters to his name, but in conformity with the equally ancient custom of attaching a certain dishonourable addition to the skirts of his jacket.

I have observed that his letter to Mr. Underwood betrayed a tincture of *Spinosism.* It may be here remarked, that during the year 1801 he composed a poem, which he arbitrarily distinguished by that appellation, singularly opposed to the tenor of the sentiments. Through the kindness of Mr. Greenough I possess a copy of it in its original state, for it was subsequently altered, and published in a collection, edited by Miss Johanna Baillie; and still more recently, it underwent farther corrections, and was printed for private circulation in the form in which I shall here introduce it.

Lo! o'er the Earth the kindling spirits pour
  The flames of life, that bounteous Nature gives;
The limpid dew becomes the rosy flower;
  The insensate dust awakes, and moves, and lives.

All speaks of change: the renovated forms
  Of long forgotten things arise again.
The light of suns, the breath of angry storms,
  The everlasting motions of the main;

These are but engines of the Eternal Will,
The One Intelligence; whose potent sway
Has ever acted, and is acting still,
Whilst stars, and worlds, and systems, all obey:

Without whose power, the whole of mortal things
Were dull, inert, an unharmonious band;
Silent as are the harp's untuned strings
Without the touches of the poet's hand.

A sacred spark, created by His breath,
The immortal mind of man his image bears;
A spirit living midst the forms of death,
Oppress'd but not subdued by mortal cares—

A germ, preparing in the winter's frost,
To rise and bud and blossom in the spring;
An unfledged eagle by the tempest tost,
Unconscious of his future strength of wing:—

The child of trial, to mortality,
And all its changeful influences given:
On the green earth decreed to move and die;
And yet by such a fate prepared for Heaven.—

Soon as it breathes, to feel the mother's form
Of orbed beauty, through its organs thrill;
To press the limbs of life with rapture warm,
And drink of transport from a living rill:

To view the skies with morning radiance bright,
Majestic mingling with the ocean blue,
Or bounded by green hills, or mountains white;
Or peopled plains of rich and varied hue:

The nobler charms astonish'd to behold
Of living loveliness, to see it move,
Cast in expression's rich and varied mould,
Awakening sympathy, compelling love:—

The heavenly balm of mutual hope to taste,
Soother of life; affection's bliss to share,
Sweet as the stream amidst the desert waste,
As the first blush of arctic daylight fair:—

To mingle with its kindred, to descry
The path of power—in public life to shine;
To gain the voice of popularity;
The idol of to-day, the man divine:—

To govern others by an influence strong
As that high law, which moves the murm'ring main;
Raising and carrying all its waves along,
Beneath the full-orb'd moon's meridian reign:—

To scan how transient is the breath of praise;
A winter's zephyr trembling on the snow,
Chill'd as it moves; or as the northern rays,
First fading in the centre, whence they flow:—

To live in forests mingled with the whole
Of natural forms, whose generations rise
In lovely change, in happy order roll
On land, in ocean, in the glittering skies:—

Their harmony to trace—The Eternal Cause
To know in love, in reverence to adore—
To bend beneath the inevitable law,
Sinking in death; its human strength no more:—

Then, as awakening from a dream of pain,
With joy its mortal feelings to resign;
Yet all its living essence to retain,
The undying energy of strength divine:

To quit the burdens of its earthly days,
To give to Nature all her borrow'd powers;
Ethereal fire to feed the solar rays,
Ethereal dew to glad the earth in showers.

The following letter was written from London, after his return from his Cornish excursion.

TO DAVIES GIDDY, ESQ.

MY DEAR SIR, Royal Institution, Nov. 14, 1801.

AFTER leaving Cornwall in August, I spent about three weeks in Bristol, and at Stowey, so that I did not arrive in London until the 20th of September.

On my arrival I found that Count Rumford had altered his plans of absence, and had left London on that very day for the Continent, purposing to return in about two months. He is now at Paris, and in about a fortnight we expect him here.

I shall soon have an opportunity of submitting Captain Trevitheck's boiler to his observation, and in my next letter I shall give you his opinion of it.

You of course have read an account of Dr. Herschel's experiments on the heat-making rays; from some late observations it appears, that there are other invisible rays beyond the violet ones, possessed of the *chemical* agency of Light. Sennebier ascertained some time ago that the violet rays blackened muriate of silver in three seconds; whereas the red rays required, for this effect, twenty minutes. Ritter and Dr. Wollaston have found that beyond the violet rays there is exerted a strong deoxidating action. Muriate of silver placed in the spectrum is not altered beyond the red rays; but it is instantly blackened when placed on the outside of the violet rays. I purpose to try whether the invisible deoxidating rays will produce light, when absorbed by solar phosphorus.

The most curious galvanic facts lately noticed, are the combustion of gold, silver, and platina. Professor Tromsdorf, by connecting the ends of a moderately strong battery with gold and silver leaf, produced the combustion of them with vivid light. In repeating the experiment on a thin slip of platina, I have produced the same effect.

I yesterday ascertained rather an important fact, namely, that a galvanic battery may be constructed *without any metallic substance!* By means of ten pieces of well-burnt charcoal, nitrous acid, and water, arranged alternately in wine-glasses, I produced all the effects usually obtained from zinc, silver, and water.

The Bakerian Lecture by Dr. Young, our Lecturer on Natural Philosophy, is now reading before the Royal Society. He attempts to revive the doctrine of Huygens and Euler, that light depends upon undulations of an ethereal medium. His proofs (i. e. his presumptive proofs) are drawn from some strong and curious analogies he has discovered between Light and Sound.

I shall strongly hope, now the Peace has arrived, to see you soon in London, as you proposed a tour through the Continent; indeed, you should fix your permanent residence in London, where alone you can do what you ought—instruct and delight numbers of improved men.

I am, my friend, yours
With unfeigned esteem and respect,
HUMPHRY DAVY.

Although during 1801 Davy had given some desultory Lectures, his splendid career cannot be said to have commenced until the following year, when on the 21st of January he delivered his Introductory Lecture to a crowded and enlightened audience in the Theatre of the Royal Institution; which was afterwards printed at the request of a respectable proportion of the Society.

It contains a masterly view of the benefits to be derived from the various branches of science. He represents the Chemist as the Ruler of all the elements that surround us, and which he employs either for the satisfaction of his wants, or the gratification of his wishes. Not contented with what is to be found on the surface of the earth, he describes him as penetrating into her bosom, and even of searching the depths of the ocean, for the purpose of allaying the restlessness of his desires, or of extending and increasing the boundaries of his power.

In examining the science of Chemistry, with regard to its great agency in the improvement of society, he offers the following almost prophetic remarks. "Unless any great physical changes should take place upon the globe, the permanency of the Arts and Sciences is rendered certain, in consequence of the diffusion of knowledge, by means of the invention of Printing; and by which those words which are the immutable instruments of thought, are become the constant and widely diffused nourishment of the mind, and the preservers of its health and energy.

"Individuals influenced by interested motives or false views, may check for a time the progress of knowledge;—moral causes may produce a momentary slumber of the public spirit;—the adoption of wild and dangerous theories, by ambitious or deluded men, may throw a temporary opprobrium on literature; but the influence of true philosophy will never be despised; the germs of improvement are sown in minds, even where they are not perceived; and sooner or later, the spring-time of their growth must arrive.

"In reasoning concerning the future hopes of the human species, we may look forward with confidence to a state of society, in which the different orders and classes of men will contribute more effectually to the support of each other than they have hitherto done. This state, indeed, seems to be approaching fast; for, in consequence of the multiplication of the means of instruction, the man of science and the manufacturer are daily becoming more assimilated to each other. The artist, who formerly affected to despise scientific principles, because he was incapable of perceiving the advantages of them, is now so far enlightened as to favour the adoption of new processes in his art, whenever they are evidently connected with a diminution of labour; and the increase of projectors, even to too great an extent, demonstrates the enthusiasm of the public mind in its search after improvement.

"The arts and sciences, also, are in a high degree cultivated and patronized by the rich and privileged orders. The guardians of civilization and of refinement, the most powerful and respected part of society, are daily growing more attentive to the realities of life,—and giving up many of their unnecessary enjoyments, in consequence of the desire to be useful, are becoming the friends and protectors of the labouring part of the community.

"The unequal division of property and of labour, the differences of rank and condition amongst mankind, are the sources of power in civilized life—its moving causes, and even its very soul. In considering and hoping that the human species is capable of becoming more enlightened and more happy, we can only expect that the different parts of the great whole of society should be intimately united together, by means of knowledge and the useful arts; that they should act as the children of one great parent, with one determinate end, so that no power may be rendered useless—no exertions thrown away.

"In this view, we do not look to distant ages, or amuse ourselves with brilliant though delusive dreams, concerning the infinite improveability of man, the annihilation of labour, disease, and even death, but we reason by analogy from simple facts, we consider only a state of human progression arising out of its present condition,—we look for a time that we may reasonably expect—FOR A BRIGHT DAY, OF WHICH WE ALREADY BEHOLD THE DAWN."

The extraordinary sensation produced amongst the members of the Institution by this first course of lectures, has been vividly described by various persons who had the good fortune to be his auditors; and foreigners have recorded in their travels the enthusiasm with which the great English chemist had inspired his countrymen.

The members of the Tepidarian Society, sanguine

in the success of their child,—for so they considered Davy,—purposely appointed their anniversary festival on the day of his anticipated triumph. They were not disappointed in their hopes; and their dinner was marked by every demonstration of hilarity. In the evening, Davy accompanied by a few friends, attended, for the first time in his life, a masquerade which was given at Ranelagh.

On the following day, he dined with Sir Harry Englefield. I have a copy of the invitation, addressed to Mr. Underwood, now before me.

DEAR UNDERWOOD,

DAVY, covered with glory, dines with me to-day at five. If you could meet him, it would give me great pleasure. Yours truly,

Tilney Street, Friday. H. C. ENGLEFIELD.

At this dinner, Sir Harry wrote a request to Davy to print his Lecture, which was signed by every one present, except Mr. Underwood, who declined, from the apprehension that the signature of so intimate a friend might give to that which was a spontaneous homage to talent, the appearance of a previously concerted scheme.

I shall here weave into my narrative some extracts from several letters, with which Mr. Purkis, one of the earliest friends of Davy, has lately favoured me.

"On his first appointment at the Royal Institution, I was specially introduced to him by a common friend, Thomas Poole, Esq. of Nether Stowey in Somersetshire; and I continued in habits of friendship with him during a great portion of his life, though somewhat less intimately during the last few years. I loved him living—I lament his early death: I shall ever honour his memory.

"The sensation created by his first course of Lectures at the Institution, and the enthusiastic admiration which they obtained, is at this period scarcely to be imagined. Men of the first rank and talent,—the literary and the scientific, the practical and the theoretical, blue-stockings, and women of fashion, the old and the young, all crowded—eagerly crowded the lecture-room. His youth, his simplicity, his natural eloquence, his chemical knowledge, his happy illustrations and well-conducted experiments, excited universal attention and unbounded applause. Compliments, invitations, and presents, were showered upon him in abundance from all quarters; his society was courted by all, and all appeared proud of his acquaintance.

"One instance of attention is particularly recalled to my memory. A talented lady, since well known in the literary world, addressed him anonymously in a poem of considerable length, replete with delicate panegyric and genuine feeling. It displayed much originality, learning, and taste: the language was elegant, the versification harmonious, the sentiments just, and the imagery highly poetical. It was accompanied with a handsome ornamental appendage for the watch, which he was requested to wear when he delivered his next lecture, as a token of having received the poem, and pardoned the freedom of the writer. It was long before the fair authoress was known to him, but they afterwards became well acquainted with each other."

I should not redeem the pledge given to my readers, nor fulfill the duties of an impartial biographer, were I to omit acknowledging that the manners and habits of Davy very shortly underwent a considerable change. Let those who have vainly sought to disparage his excellence, enjoy the triumph of knowing that he was not perfect; but it may be asked in candour, where is the man of twenty-two years of age, unless the temperature of his blood were below zero, and his temperament as dull and passionless as the fabled god of the Brahmins, who could remain uninfluenced by such an elevation? Look at Davy in the laboratory at Bristol, pursuing with eager industry various abstract points of research; mixing only with a few philosophers, sanguine like himself in the investigation of chemical phenomena, but whose sphere of observation must have been confined to themselves, and whose worldly knowledge could scarcely have extended beyond the precincts of the Institution in which they were engaged. Shift the scene—behold him in the Theatre of the Royal Institution, surrounded by an aristocracy of intellect as well as of rank; by the flowers of genius, the *élite* of fashion, and the beauty of England, whose very respirations were suspended in eager expectation to catch his novel and satisfactory elucidations of the mysteries of Nature. Could the author of the Rambler have revisited us, he would certainly have rescinded the passage in which he says—"All appearance of science is hateful to women; and he who desires to be well received by them, must qualify himself by a total rejection of all that is rational and important; must consider learning as perpetually interdicted, and devote all his attention to trifle, and all his eloquence to compliment."

It is admitted that his vanity was excited, and his ambition raised, by such extraordinary demonstrations of devotion; that the bloom of his simplicity was dulled by the breath of adulation; and that, losing much of the native frankness which constituted the great charm of his character, he unfortunately assumed the garb and airs of a man of fashion; let us not wonder if, under such circumstances, the inappropriate robe should not always have fallen in graceful draperies.

At length, so popular did he become, under the auspices of the Duchess of Gordon and other leaders of high fashion, that even their *soirées* were considered incomplete without his presence; and yet these fascinations, strong as they must have been, never tempted him from his allegiance to Science: never did the charms of the saloon allure him from the pursuits of the laboratory, or distract him from the duties of the lecture-room. The crowds that repaired to the Institution in the morning were, day after day, gratified by newly devised and highly illustrative experiments, conducted with the utmost address, and explained in language at once perspicuous and eloquent.

He brought down Science from those heights which were before accessible only to a few, and placed her within the reach of all; he divested the goddess of all severity of aspect, and represented her as attired by the Graces.

It is perhaps not possible to convey a better idea of the fascination of his style, than by the relation of the following anecdote. A person having observed the constancy with which Mr. Coleridge attended these lectures, was induced to ask the poet what attractions he could find in a study so unconnected with his known pursuits. "I attend Davy's lectures," he replied, "to increase my stock of metaphors."

But, as Johnson says, in the most general applause some discordant voices will always be heard; and so was it upon the present occasion. It was urged by several modern *Zoili*, that the style was far too florid and imaginative for communicating the plain lessons of truth; that he described objects of Natural History by inappropriate imagery, and that violent conceits frequently usurped the place of philosophical definitions. This was Bœotian criticism; the Attic spirits selected other points of attack: they rallied him on the ground of affectation, and whimsically represented him as swayed by a mawkish sensibility, which constantly betrayed him into absurdity. There might be some show of justice in this accusation: The world was not large enough to satisfy the vulgar ambition of the conqueror, but the minutest production of nature afforded ample range for the scrutinizing intelligence of the philosopher; and he would consider a particle of crystal with so delicate a regard for its minute beauties, and expatiate with so tender a tone of interest on its fair proportions, as almost to convey an idea that he bewailed the condition of necessity which for ever allotted it so slender a place in the vast scheme of creation.

After the observations which have been offered with regard to the injurious tendency of metaphors in all matters relating to science, I may probably be charged with inconsistency in defending Davy from the attacks thus levelled against his style. We need not the critic to remind us that the statue of a Lysippus may be spoiled by gilding; but I would observe that the style which cannot be tolerated in a philosophical essay, may under peculiar circumstances be not only admissible, but even expedient, in a popular lecture. "*Neque ideo minus efficaces sunt orationes nostræ quia ad aures judicantium cum voluptate perveniunt. Quid enim si infirmiora horum temporum templa credas, quia non rudi cæmento, et informibus tegulis exstruuntur; sed marmore nitent et auro radiantur?*"

Let us consider, for a moment, the class of persons to whom Davy addressed himself. Were they students prepared to toil with systematic precision, in order to obtain knowledge as a matter of necessity?—No—they were composed of the gay and the idle, who could only be tempted to admit instruction by the prospect of receiving pleasure,—they were children, who could only be induced to swallow the salutary draught by the honey around the rim of the cup.

It has been well observed, that necessity alone can urge the traveller over barren heaths and snow-topped mountains, while he treads with rapture along the fertile vales of those happier climes where every breeze is perfume, and every scene a picture.

If Science can be promoted by increasing the number of its votaries, and by enlisting into its service those whom wealth and power may render valuable as examples or patrons, there does not exist a class of philosophers to which we are more largely indebted than to popular lecturers, or to those whose eloquence has clothed with interest, subjects otherwise severe and uninviting. How many disciples did Mineralogy acquire through the lectures of Dr. Clarke at Cambridge, who may truly be said to have covered a desert with verdure, and to have raised from barren rocks flowers of every hue and fragrance! In the sister university, what an accession of strength and spirit have the animated discourses of Dr. Buckland brought to the ranks of Geologists! To judge fairly of the influence of a popular style, we should acquaint ourselves with the effects of an opposite method; and if an appeal be made to experience, I may very safely abide the issue. Dr. Young, whose profound knowledge of the subjects he taught, no one will venture to question, lectured in the same theatre, and to an audiences imilarly constituted to that which was attracted by Davy, but he found the number of his attendants diminish daily, and for no other reason than that he adopted too severe and didactic a style.*

* From the following minute it would appear, that Dr. Young's connection with the Royal Institution was but of short duration. It will be remembered that his appointment took place on July 6, 1801.

"Resolved—

In speaking of Davy's lectures as mere specimens of happy oratory, we do injustice to the philosopher. Had he merely added the Corinthian foliage to a temple built by other hands, we might have commended his taste, and admired his talent of adaptation, and there our eulogium must have ended; but the edifice itself was his own; he dug the materials from the quarry, formed them into a regular pile, and then with his masterly touch added to its strength beauty, and to its utility grace.

In addition to these morning lectures, we find that he was also engaged in delivering a course in the evening; of which the following notice is extracted from one of the scientific journals of the time. "On the 25th, Mr. Davy commenced a course of lectures on Galvanic phenomena. Sir Joseph Banks, Count Rumford, and other distinguished philosophers, were present. The audience were highly gratified, and testified their satisfaction by general and repeated applause.

"Mr. Davy, who appears to be very young, acquitted himself admirably. From the sparkling intelligence of his eye, his animated manner, and the *tout ensemble*, we have no doubt of his attaining a very distinguished eminence."

From a Minute entered on the Records of the Institution, it appears that, at a meeting of Managers held on the 31st of May 1802, it was moved by Sir Joseph Banks, and seconded by Mr. Sullivan,—

"That Mr. Humphry Davy be for the future

"Resolved—That Dr. Young be paid the balance of two years' complete salary, and that his engagement with the Institution terminate from this time—July 4, 1803."

styled *Professor of Chemistry* to the Royal Institution."

A sufficient proof of the universal feeling of admiration which his lectures had excited.

The success of his exertions is communicated by him to his early friend, in the following letter.

TO DAVIES GIDDY, ESQ.

DEAR FRIEND,

SINCE the commencement of the Session at the Institution, I have had but few moments of leisure. The composition of a first course of lectures, and the preparation for experiments, have fully occupied my time; and the anxieties and hopes connected with a new occupation have prevented me from paying sufficient attention even to the common duties and affections of life. Under such circumstances, I trust you will pardon me for having suffered your letters to remain so long unanswered. In human affairs, anticipation often constitutes happiness: your correspondence is to me a real source of pleasure, and, believe me, I would suffer no opportunity to escape of making it more frequent and regular.

My labours in the Theatre of the Royal Institution have been more successful than I could have hoped from the nature of them. In lectures, the effect produced upon the mind is generally transitory; for the most part, they amuse rather than instruct, and stimulate to enquiry rather than give information. My audience has often amounted to four and five hundred, and upwards; and amongst them some promise to become permanently attached to Chemistry. This science is much the fashion of the day.

Amongst the latest scientific novelties, the two new planets occupy the attention of Astronomers, while Natural Philosophers and Chemists are still employed upon Galvanism.

In a paper lately read before the Royal Society, Dr. Herschel examines the magnitudes of the bodies discovered by Mr. Piazzi and Dr. Olbers. He supposes the apparent diameter of Ceres to be about 22″, and that of Pallas, 17″ or 13″, so that their real diameters are 163, and 95 or 71 English miles—How small! The Doctor thinks that they differ from planets in their general character, as to their diminutive size, the great inclination of their orbits, the coma surrounding them, and as to the proximity of their orbits.—From comets, in their want of their eccentricity, and any considerable nebulosity. He proposes to call them *Asteroids*.

I mentioned to you in a former letter the great powers of Galvanism in effecting the combustion of metals. I have lately had constructed for the laboratory of the Institution, a battery of immense size: it consists of four hundred plates of five inches in diameter, and forty, of a foot in diameter. By means of it, I have been enabled to inflame cotton, sulphur, resin, oil, and ether; it fuses platina wire, and makes red hot and burns several inches of iron wire of 1-300th of an inch in diameter; it easily causes fluid substances, such as oil and water, to boil, decompounds them, and converts them into gases. I am now examining the agencies of it upon certain substances that have not as yet been decomposed, and in my next letter I hope to be able to give you an account of my experiments.

I shall hope soon to hear that the roads of England are the haunts of Captain Trevitheck's dragons. You have given them a characteristic name.

I wish any thing would happen to tempt you to visit London. You would find a number of persons very glad to see you, with whose attentions you could not be displeased. With unfeigned respect,

Yours sincerely,

H. Davy.

It is perhaps not possible to imagine a greater contrast, than between the elegant manner in which Davy conducted his experiments in the theatre, and the apparently careless and slovenly style of his manipulations in the laboratory: but in the one case he was communicating knowledge; in the other, obtaining it. Mr. Purkis relates an anecdote very characteristic of this want of refinement in his working habits. "On one occasion, while reading over to me an introductory lecture, and wishing to expunge a needless epithet, instead of taking up the pen, he dipped his forefinger into the ink-bottle, and thus blotted out the unmeaning expletive."

It was his habit in the laboratory, to carry on several unconnected experiments at the same time, and he would pass from one to the other without any obvious design or order: upon these occasions he was perfectly reckless of his apparatus; breaking and destroying a part, in order to meet some want of the moment. So rapid were all his movements, that, while a spectator imagined he was merely making preparations for an experiment, he was actually obtaining the results, which were just as accurate as if a much longer time had been expended. With Davy, rapidity was power.

The rapid performance of intellectual operations was a talent which displayed itself at every period of his life. We have heard with what extraordinary rapidity he read at the age of five years; and we now learn that his chemical enquiries were conducted with similar facility and quickness.

His early friend Mr. Poole bears his testimony to the existence of the same quality in the following passage, extracted from a letter I had lately the favour of receiving from him. "From my earliest knowledge of my admirable friend, I consider his most striking characteristic to have been the quickness and truth of his apprehension. It was a power of reasoning so rapid, when applied to any subject, that he could hardly be himself conscious of the process; and it must, I think, have been felt by him, as it appeared to me, pure intuition. I used to say to him, 'You understand me before I half understand myself?'

"I recollect on our first acquaintance, he knew but little of the practice of agriculture. I was at that time a considerable farmer, and very fond of the occupation. During his visits in those days, I was at first something like his teacher, but my pupil soon became my master both in theory and practice."

The chemical manipulations of Wollaston and Davy offered a singular contrast to each other, and might be considered as highly characteristic of the temperaments and intellectual qualities of these remarkable men. Every process of the former was regulated with the most scrupulous regard to microscopic accuracy, and performed with the utmost neatness of detail. It has been already stated with what turbulence and apparent confusion the experiments of the latter were conducted; and yet each was equally excellent in his own style; and, as artists, they have not unaptly been compared to Teniers and Michael Angelo. By long discipline, Wollaston had acquired such power in commanding and fixing his attention upon minute objects, that he was able to recognise resemblances, and to distinguish differences, between precipitates produced by re-agents, which were invisible to ordinary observers, and which enabled him to submit to analysis the minutest particle of matter with success. Davy, on the other hand, obtained his results by an intellectual process, which may be said to have consisted in the extreme rapidity with which he seized upon, and applied, appropriate means at appropriate moments.

Many anecdotes might be related in illustration of the curiously different structure of the minds of these two ornaments of British Science. The reader will, in the course of these memoirs, be furnished with sufficient evidence of the existence of those qualities which I have assigned to Davy; another biographer will no doubt ably illustrate those of Dr. Wollaston.

I shall only observe, that to this faculty of minute observation, which Dr. Wollaston applied with so much advantage, the chemical world is indebted for the introduction of more simple methods of experimenting, — for the substitution of a few glass tubes, and plates of glass, for capacious retorts and receivers, and for the art of making grains give the results which previously required pounds. A foreign philosopher once called upon Dr. Wollaston with letters of introduction, and expressed an anxious desire to see his laboratory. "Certainly," he replied; and immediately produced a small tray containing some glass tubes, a blow-pipe, two or three watch-glasses, a slip of platinum, and a few test bottles.

Wollaston appeared to take great delight in showing by what small means he could produce great results. Shortly after he had inspected the grand galvanic battery constructed by Mr. Children, and had witnessed some of those brilliant phenomena of combustion which its powers produced, he accidentally met a brother chemist in the street, and seizing his button, (his constant habit when speaking on any subject of interest,) he led him into a secluded corner; when taking from his waistcoat pocket a tailor's thimble, which contained a galvanic arrangement, and pouring into it the contents of a small phial, he instantly heated a platinum wire to a white heat.

There was another peculiarity connected with Wollaston's habit of minute observation: it enabled him to press into his service, at the moment, such ordinary and familiar materials as would never have occurred to less observing chemists. Mr. Brande relates an anecdote admirably calculated to exemplify this habit. He had called upon Dr. Wollaston to consult him upon the subject of a calculus;

—it will be remembered that neither phosphate of lime, constituting the '*bone earth*' species, nor the ammoniaco-magnesian phosphate, commonly called the '*triple phosphate*,' is *per se* fusible; but that when mixed, these constitute the '*fusible calculus*' which readily melts before the blow-pipe. Dr. Wollaston, on finding the substance under examination refractory, took up his paper-folder, and scraping off a fragment of the ivory, placed it on the specimen, when it instantly fused.*

Having contrasted the manipulations of Davy, as exhibited in the theatre, with those performed by him in the laboratory, it may, in this place, be interesting to offer a few remarks upon the difference of his style as a lecturer and as a writer. Whatever diversity of opinion may have been entertained as to the former, I believe there never was but one sentiment with respect to the latter. There is an ethereal clearness of style, a simplicity of language, and, above all, a freedom from technical expression, which render his philosophical memoirs fit studies and models for all future chemists. Mr. Brande, in a late lecture delivered before the members of the Royal Institution, very justly alluded to this latter quality of his writings, and forcibly contrasted it with the system of Berzelius, of whom it is painful to speak but in terms of the most profound respect, and yet it is impossible not to express a deep regret at this distinguished

* I have lately been informed that the idea of constructing an instrument like the Camera Lucida, first suggested itself to Dr. Wollaston, on his noticing certain phenomena occasioned by a crack in the glass before which he was shaving himself.

chemist's introduction of a system of technical expressions, which from its obscurity is calculated to multiply rather than to correct error, and from its complications, to require more labour than the science to which it administers: to apply the quaint metaphor of Locke, "it is no more suited to improve the understanding, than the move of a jack is to fill our bellies."

From the readiness with which some continental chemists have adopted such terms, and from the spirit in which they have defended them, one might almost be led to suspect that they believed them, like the words used by the Magi of Persia, to possess a cabalistic power. Davy foresaw the injury which science must sustain from such a practice, and endeavoured, both by precept and example, to discountenance it.

With regard to the introduction of a figurative and ornamental style into memoirs purely scientific, no one could entertain a more decided objection; and in his "Last Days," he warns us against the practice.

"In detailing the results of experiments, and in giving them to the world, the chemical philosopher should adopt the simplest style and manner; he will avoid all ornaments, as something injurious to his subject, and should bear in mind the saying of the first king of Great Britain, respecting a sermon which was excellent in doctrine, but overcharged with poetical allusions and figurative language,— "that the tropes and metaphors of the speaker were like the brilliant wild flowers in a field of corn, very pretty, but which did very much hurt the corn."

CHAPTER IV.

Davy makes a tour with Mr. Purkis, through Wales.—Beautiful phenomenon observed from the summit of Arran Benllyn.—Letter to Mr. Gilbert.—Journal of the Institution.—Davy's papers on Eudiometry, and other subjects.—His first communication to the Royal Society, on a new galvanic pile.—He is proposed as a Fellow, and elected into the Society.—His paper on astringent vegetable substances, and on their operation in tanning leather. —His letter to Mr. Poole.—He is appointed Chemical Lecturer to the Board of Agriculture.—He forms friendships with the Duke of Bedford, Mr. Coke, and many other celebrated agriculturists. —Attends the sheep-shearing at Holkham and Woburn.—Composes a Prologue to the "Honey-Moon."

After the fatigues and anxieties of his first session, Davy sought relaxation and repose amidst the magnificent scenery of Wales. The following letter will serve more fully to exhibit the enthusiasm he experienced in contemplating Nature in her wild and simplest forms.

TO SAMUEL PURKIS, ESQ.

MY DEAR FRIEND, Matlock, August 15, 1802.

Had I been alone, and perfectly independent as to my plans, I should probably have written to you long ago. I should have begged you to hasten your departure, so that we might have rejoiced together in Wales, under the influence of that moon which is now full in all its glory; for Derbyshire, taken as a whole, has not pleased me. A few beautiful valleys, placed at the distance of many miles from each other, do not compensate for the almost uniform wildness and brown barrenness of the hills and plains; and in the watering places, there is little amongst the *living beings* to awaken deep moral feelings, or to call the nobler powers of the mind, which act in consequence of sympathy, into existence.

I have longed for the mountain scenery, and for the free inhabitants of North Wales; and even the majestic valleys of the Wye and the Derwent have been to me but typical of something more perfect in beauty and grandeur.

Whenever it shall seem fitting to you, I shall be prepared for our long contemplated journey, and do not delay your departure;—*before* the 21st would be more agreeable to me than *after* that period; and then we shall be able to view the horns of the next moon, where they are most beautiful.

I have enquired much concerning Dove-dale, since I have been here, and, from the most accurate accounts, I am inclined to believe that it is inferior, in point of sublime scenery, to *Chee Tor*, near Bakewell, and in beauty, to the valley of the *Great Tor*, in which I am now writing. On the whole, I think your best plan will be to meet me at Matlock, which you must see, and then, in our route to Buxton, we can visit the valley of the Wye, and the most noble *Chee Tor*.

Concerning the excursion of Dove-dale, I am

undecided, and it shall depend upon you to determine with regard to it.

As one great object in our excursions is to view Nature and man in their most simple forms, and to gain a temporary life of new impressions, I submit to you whether it will not be best to steer clear of *towns, cities,* and *civilized society,* in which, for the most part, we can see what we have only seen before.

If we visit Sir Joseph Banks, it certainly should be only *en passant:* and to see that most excellent personage, and to be obliged to quit him immediately, will be at least painful; for the respectful feelings he produces in the mind are always modified by affection.

I have no room to give you the quantity of information that I have gained concerning the places and people of Wales; this shall serve for our Derbyshire chat. I thank you much for your last kind letter, which gave me high pleasure. You possess the true spirit of composition, which embodies facts in words. I am, &c.

H. Davy.

I am informed by Mr. Purkis, that in the latter end of this summer, he made a pedestrian tour with Davy, through North and South Wales, and he has transmitted to me the following account of this excursion.—"We visited every place possessing any remains of antiquity, any curious productions of nature or art, and every spot distinguished by romantic and picturesque scenery. Our friend's diversified talents, with his knowledge of Geology, and

Natural History in general, rendered him a most delightful companion in a tour of this description. Every mountain we beheld, and every river we crossed, afforded a fruitful theme for his scientific remarks. The form and position of the mountain, with the several *strata* of which it was composed, always procured for me information as to its character and classification; and every bridge we crossed invariably occasioned a temporary halt, with some appropriate observations on the productions of the river, and on the diversion of angling.

"In one of our morning excursions in North Wales, we ascended the summit of *Arran Benllyn*, a celebrated mountain, inferior only to Snowdon and Cader Idris, a few miles from the lake of Bala. Here we were fortunate in beholding a scene of extraordinary sublimity, seldom witnessed in this climate. From the top of this mountain we looked down, about mid-day, on a deep valley eight or ten miles in length, and as many in breadth, the whole of which, for a considerable height from the surface of the ground, was filled with beautiful clouds, while the atmosphere around the summit on which we stood was perfectly clear, and the sky above us of a deep blue colour. The clouds in the valley were in irregular, gentle undulations, dense, compact, and continuous, of that kind which is denominated by Meteorologists *cirro-cumulus*, and by the vulgar, *woolpack* clouds, such as are often seen in the higher regions on a fine summer's day. The sun shone with great brilliancy, and illumined their various forms with silver, grey and blue tints of exquisite beauty As there was scarcely a breath of air stirring below the mountain, this aggregation of clouds, probably occasioned by some electrical agency, remained fixed and stationary, as if identified with the valley. The higher parts of most of the surrounding hills were enveloped in mist, above which the tops of *Snowdon* and *Cader Idris* towered distinctly visible, and appeared like small islands rising out of the sea. This scene altogether was one of inexpressible magnificence and grandeur, filling the mind with awe and rapture. We seemed to feel ourselves like beings of a higher order in a celestial region, looking down on the lower world with conscious superiority.

"After sitting and ruminating on this sublime spectacle for two or three hours, we left the summit of the mountain with reluctance, and, slowly descending, rested at intervals, and often cast a longing, lingering look behind.

"On reaching our comfortable inn at Bala, while waiting for dinner, Davy walked about the room, and, as if by inspiration, broke out in a beautiful impassioned apostrophe on the striking scene we had so recently witnessed. It was in a kind of unmeasured blank verse, highly animated and descriptive, at once poetical and philosophical. At the conclusion of this eloquent effusion, I endeavoured to recollect and commit it to writing, but I could not succeed, and Davy was too modest to assist my memory.

"In a tour through North Wales, where the few small inns have seldom any spare rooms, different parties are often obliged to sit in the same apartment, and to eat at the same table. Hence we were occasionally introduced to characters of various descriptions, some of whom gratified us by their agreeable qualities, while others disgusted us by their ignorance and impertinence. On one occasion, after a heavy shower of rain in the middle of August, we were drying our clothes by the fire in the little Inn at *Tan y Bwlch*, when the landlord requested us to admit a gentleman, who was very wet. A young man, of pleasing appearance and manner, was then introduced, and after some common-place observations, we sat down to dinner. The stranger was evidently a man of education and acquirements, and after the cloth had been removed, he began to discourse very fluently on scientific subjects. He talked of oxygen and hydrogen, of hornblende, and the *Grawacké* of Werner, and geologists, in the most familiar tone of self-complacency.

"Davy's youth, simplicity of manner, and cautious concealment of superior knowledge, not exciting constraint, our companion was naturally induced to deliver his opinions with the utmost freedom and confidence on all subjects. We commenced on poetry and painting; the sublime and beautiful; then proceeded to mineralogy, astronomy, &c. and occasionally digressed on topics of mirth and humour, so that the evening was passed with general satisfaction.

"When Davy had retired to rest, and I was left alone with our companion, I enquired how he liked my friend, and whether he considered him a proficient in science, and versed in chemistry and geology? He answered coolly, that 'he appeared to be rather a clever young man, with some general scientific knowledge.' He then asked his name, and when I announced 'Davy, of the Royal Institution,' the stranger seemed thunderstruck, and exclaimed, 'Good God! was that really Davy? How have I exposed my ignorance and presumption!' It is scarcely necessary to add, that at the breakfast table the next morning, he talked on subjects of science with less volubility than on the preceding evening."

After Davy's return from this expedition, he wrote the following letter.

TO DAVIES GIDDY, ESQ.

Royal Institution, Oct. 26, 1802.

MY DEAR SIR,

It is long since I have had the pleasure of hearing from you. You probably received a hasty letter that I wrote to you in the beginning of the summer. Since that period, I have been idling away much of my time in Derbyshire and North Wales.

Till very lately, I had hopes of being able to spend a few weeks of the autumn in Cornwall, but I now find that it will not suit with my occupations. Not having it in my power to see you, you may believe that I am most anxious to hear from you.

We hear, at this time, in London of comparatively few scientific novelties. The wonders of revolutionized Paris occupy many of our scientific men; and the summer and autumn are not the working seasons in great cities. The rich and fashionable part of the community think it their duty to kill time in the country, and even philosophers are more or less influenced by the spirit of the times.

In the last volume of the Manchester Memoirs, *i. e.* the fifth, are some papers of Mr. Dalton on the Constitution of the Atmosphere; on the expansive powers of Steam; and on the dilatation of Elastic Fluids by Heat. As far as I can understand his subjects, the author appears to me to have executed them in a very masterly way. I wish very much to have your judgment upon his opinions, some of which are new and very singular.

Have you yet seen the theory of my colleague, Dr. Young, on the undulations of an Ethereal Medium as the cause of Light? It is not likely to be a popular hypothesis, after what has been said by Newton concerning it. He would be very much flattered if you could offer any observations upon it, whether for or against it. The paper is in the last volume of the Transactions.

I believe I mentioned to you in a former letter that *Terra Japonica*, or *Extractum Catechu*, contained a very large proportion of the tanning principle. My friend Mr. Purkis, an excellent practical tanner, has lately tried some experiments upon it in the large way. It answers very well, and I am now wearing a pair of shoes, the leather of one of which was tanned with oak-bark, and that of the other with *Terra Japonica*; and they appear to be equally good. We are in great hopes that the East India Company will consent to the importation of this article. One pound of it goes at least as far as nine pounds of oak-bark; and it could certainly be rendered in England for less than four-pence the pound: oak-bark is nearly one penny per pound.

The *Zoonic acid,* which M. Berthollet supposed to be a peculiar acid, has been lately shown by M. Thenard to be only acetous acid, holding a peculiar animal matter in solution.

Gregory Watt is just returned from the Continent, where he has passed the last fifteen months. He has been much delighted with his excursion, but his health is at present bad. I trust, however, that English roast-beef and English customs will speedily restore it.

We are publishing, at the Royal Institution, a Journal of Science, which contains chiefly abridged accounts of what is going on in different parts of Europe, with some original papers; and in hopes that its diffusion may become more general, we have fixed its price at one shilling. As soon as I have an opportunity, I will send you the last numbers of it.

I am beginning to think of my Course of Lectures for the winter. In addition to the common course of the Institution, I have to deliver a few lectures on Vegetable Substances, and on the connection of Chemistry with Vegetable Physiology, before the Board of Agriculture. I have begun some experiments on the powers of soils to absorb moisture, as connected with their fertility. I have, for this purpose, made a small collection of those of the *calcareous* and *secondary* countries, and I wish very much for some specimens from the *granitic* and *schistose* hills of Cornwall. If you could, without much trouble, cause to be procured from your estates different pieces of uncultivated soil, of about a pound weight each, I should feel much obliged to you. They should be accompanied by specimens of the stone or strata on which they lie.

I am, dear Sir, with affection and respect, yours,

H. DAVY.

Of the Journals alluded to in the above letter, it would seem that Davy and Dr. Young were the joint editors. The former appears both as a reviewer and an original writer, and in each capacity we recognise the peculiarities of his genius: in the one case, by the quickness with which he detects error; and in the other, by the avidity with which he apprehends truth.

It will not be uninteresting to take a short review of his original communications, especially as the work has become extremely scarce; indeed, as it was published in numbers, it is very probable that only a few copies have escaped the common fate of periodicals.

His first paper is entitled, "An Account of a New Eudiometer," and has for its principal object the recommendation of the solution of the green muriate, or sulphate of iron, impregnated with *nitrous* gas; the knowledge of the properties of which, in absorbing oxygen gas, arose out of those experiments to which an allusion has been already made.*

This test is prepared by transmitting a current of nitrous gas through a saturated solution of the salt

* See page 88.

of iron. As the absorption of the gas proceeds, the solution acquires the colour of a deep olive brown; and when the impregnation is completed, it appears opaque and almost black. The process is apparently owing to a simple elective attraction; in no case is the gas decomposed; and under the exhausted receiver it resumes its elastic form, leaving the fluid with which it was combined, unaltered in its properties. The test, therefore, can only be regarded as a convenient modification of that of Priestley, in which the nitrous gas was presented to the atmospheric air to be examined, without the intervention of any third body.

The only apparatus required for the application of the test, as suggested by Davy, is a small graduated tube, having its capacity divided into one hundred parts, and a vessel for containing the fluid. The tube, after being filled with the air to be examined, is introduced into the solution, and shaken in contact with it; when the air will be rapidly diminished in volume, and the whole of its oxygen, in a few minutes, condensed into nitrous acid.

By means of this test, Davy informs us that he examined the atmosphere in different places, without being able to ascertain any notable difference in the proportions of its component parts.

Twenty-eight years have elapsed since the publication of this paper; and yet, amidst the rapid progress of discovery, Eudiometry has not been able even to modify the results it has given us; but the reader will be pleased to remember, that by these tests it is only professed to show the relative proportions of oxygen in air; the salubrity of an atmosphere depends upon many other causes, especially its condition with regard to moisture, which, in a variety of ways, exerts an influence upon the structures of the body.

In this Journal we also find several original communications from Davy on galvanic phenomena, which will be noticed on a future occasion. There is likewise a paper of considerable interest, entitled, "An Account of a method of copying Paintings upon Glass, and of making Profiles, by the agency of Light upon Nitrate of Silver, invented by T Wedgwood, Esq.: with Observations by H. Davy."

In the first place, he gives an account of the experiments of Mr. Wedgwood, and then, with his usual sagacity, extends our knowledge of the subject by his own researches.

Chemists had been long acquainted with the fact, that white paper, or white leather, moistened with a solution of the *nitrate of silver*, although it does not undergo any change when kept in a dark place, will speedily change colour on being exposed to daylight; and that, after passing through different shades of grey and brown, it will at length become nearly black. These alterations in colour take place more speedily in proportion as the light is more intense. In the direct beams of the sun, two or three minutes are sufficient,—in the shade, several hours are required, to produce the full effect; and light transmitted through differently coloured glass, acts upon it with different degrees of intensity. It is found, for instance, that red rays, or the common sunbeams passed through red glass, have very little action upon it; yellow and green are more efficacious; but blue and violet produce the most decided and powerful effects. Davy observes that these facts were analogous to those which were long ago observed by Scheele, and confirmed by Senebier.

To Mr. Wedgwood, however, belongs the merit of having first applied them for the ingenious purpose of copying engravings, &c. His first attempt was to copy the images formed by the *camera obscura;* but they were found to be too faint to produce, in any moderate time, the necessary changes upon the *nitrate of silver.* With paintings on glass he was more successful; for the copying of which, the solution should be applied on leather, which is more readily acted upon than paper. When a surface thus prepared is placed behind a painting on glass, exposed to the solar light, the rays transmitted through the differently coloured surfaces produce distinct tints of brown or black, sensibly differing in intensity, according to the shades of the picture; and where the light is unaltered, the colour of the *nitrate* becomes deepest.

Besides this application of the method of copying, there are many others. It may be rendered subservient for making delineations of all such objects as are possessed of a texture partly opaque, and partly transparent; such, for instance, as the woody fibres of leaves, and the wings of insects; for which purpose, it is only necessary to cause the direct solar light to pass through them, and to receive the shadows upon prepared leather.

To Davy we are indebted for an extremely beautiful application of this principle,—that of copying small objects produced by means of the solar microscope. For the success, however, of this experiment, it is necessary that the prepared leather should be placed at a small distance only from the lens.

The copy of a painting, or the profile of an object, thus obtained, must of course be preserved in an obscure place; for all the attempts that have been made to prevent the uncoloured parts of the copy from being acted upon by light, have hitherto been unsuccessful. They have been covered with a thin coating of fine varnish; and they have been submitted to frequent washings; yet, even after this latter operation, it would seem that a sufficient quantity of the active matter will still adhere to the white parts of the surface, and cause them to become dark on exposure to the rays of the sun. From this circumstance, Davy thinks it probable that a portion of the metallic oxide abandons its acid, to enter into union with the animal or vegetable substance, so as to form with it an insoluble compound.

It will be remembered that Davy had made some early experiments on the collision of flint and steel *in vacuo*:* we find in the Royal Institution Journal a farther investigation of the subject; when he admits that, although sparks are not produced under these circumstances, yet that a faint light becomes visible. In many instances, he refers the phenomenon to electricity excited by friction, more especially in the instances of glass, quartz, sugar, &c. which give out light when rubbed. In other cases, he considers it probable that a species of phosphores-

* See page 67.

cence may be occasioned by the heat; and he thinks that there may occasionally take place an actual ignition of abraded particles, in consequence of their imperfect conducting power: a supposition which he thinks receives strong support from an experiment of Mr. Wedgwood, who found that a piece of window-glass, when brought into contact with a revolving wheel of grit, became red hot at its point of friction, and gave off luminous particles that were capable of inflaming gunpowder and hydrogen gas.

We shall also find in this volume an account of some observations which he made upon the motions of small pieces of *acetate of potash*, during their solution, upon the surface of water. After the interesting and extraordinary observations of Mr. Brown, every phenomenon of this kind is calculated to excite attention.

Davy states that the fragments were agitated by very singular motions during the time of their solution, sometimes revolving for a second or two, and then moving rapidly backwards and forwards in various directions. He considers the phenomenon as evidently connected with the rapid process of solution, since the motions became weaker as the point of saturation approached. The thinnest film of oil, or of ether, wholly destroyed the effect. Those pieces which were most irregular in their forms underwent, by far, the most rapid motions; from which, he thinks, it would appear, that the phenomenon was in some measure owing to changes in the centre of gravity of the particles during their solution. The projectile motions, however, would seem to depend upon the continual descent of a current of the saline solution from the agitated particle, in consequence of which, the surrounding water would press upon different parts of it with different degrees of force. Besides which, an increase of temperature, which was found to accompany the solution of the salt, might in a degree modify the effect.

His first communication to the Royal Society was entitled "An Account of some Galvanic Combinations, formed by an arrangement of single metallic plates and fluids, analogous to the Galvanic Apparatus of M. Volta."

It was read on the 18th of June 1801, and will be examined in a future part of these memoirs.

The certificate, recommending him as a candidate for the honour of a seat in the Royal Society, was read for the first time on the 21st of April 1803; and having been duly suspended in the meeting-room, during ten sittings of the Society, according to the statute, he was put to the ballot, and elected on the 17th of November in the same year.

As every circumstance connected with the progress of Davy will be hereafter viewed with considerable interest, I shall here introduce the form of the certificate, and record the names of those Fellows who sanctioned it by their signatures.

"Humphry Davy, Esq. Professor of Chemistry in the Royal Institution of Great Britain, a gentleman of very considerable scientific knowledge, and author of a paper in the Philosophical Transactions, being desirous of becoming a Fellow of the Royal

Society, we the undersigned do from our personal knowledge recommend him as deserving that honour, and as likely to prove an useful and valuable member.

(Signed)	MORTON,	W. G. JORDAN,
	R. J. SULLIVAN,	JOHN WALKER,
	KINNAIRD,	RICHARD CHENEVIX,
	CHARLES HATCHETT,	ALEXANDER CRICHTON,
	THOMAS YOUNG,	HENRY C. ENGLEFIELD,
	WEBB SEYMOUR,	CHARLES WILKINS,
	W. G. MATON,	GIFFIN WILSON,
	THOMAS RACKETT,	GILBERT BLANE,
	JAMES EDWARD SMITH,	EDWARD FORSTER."

On the 7th of July, in the same year, he was elected an Honorary Member of the Dublin Society, having been proposed from the chair by the Vice-President, General Vallancey.

It has been stated that, shortly after Davy's arrival at the Institution, the Managers, being anxious to encourage all investigations of a practical tendency, directed him to deliver a series of lectures on the art of tanning. With this view, he entered into a scientific examination of the subject, in which he was encouraged by Sir Joseph Banks, the liberal patron and promoter of all useful knowledge, who supplied him with various materials for experiment. The subject had recently attracted considerable attention, both at home and abroad, but much still remained to be effected; and Davy succeeded in adding many important facts to the general store.

In the Royal Institution Journal already noticed, we find several communications from him, under the titles of "Observations on different methods of obtaining Gallic Acid;"—"On the processes of Tanning," &c. All the new facts however, discovered in the course of his experiments, were embodied in a long and elaborate memoir, which was read before the Royal Society on the 24th of February 1803, and published in the Philosophical Transactions for that year. It was entitled "An Account of some Experiments and Observations on the constituent parts of certain astringent Vegetables, and on their operation in Tanning. By Humphry Davy, Professor of Chemistry in the Royal Institution. Communicated by the Right Honourable Sir Joseph Banks, P. R. S."

Although Seguin and Proust had already examined many of the properties of that vegetable principle to which the name of *tannin* had been given, yet its affinities had been but little examined; and the manner in which its action upon animal matters may be modified by combination with other substances, had been still less considered.

His principal design in this enquiry was to elucidate the practical part of the art of tanning skins, so as to form leather; but in pursuing this object, he was necessarily led into chemical investigations connected with the analysis of the various bodies containing the tanning principle, and the peculiar properties and value of each.

The vegetable principles that had been regarded as more usually present in astringent infusions, are *tannin*, *gallic acid*, and *extractive matter*. In attempting to ascertain the relative proportions of *tannin* contained in different infusions, Davy was led, after various trials, to prefer the generally received method of precipitating by means of *gelatine* procured from isinglass. In using this test, however, he discovered that several precautions were necessary;—that the solution should be quite fresh,—that it should be as much saturated as may be compatible with its fluidity,—and that the precipitate obtained should be reduced to a uniform degree of dryness.

It is evident that if the quantity of gelatine in the solution, employed as the precipitant, be known, it will only be necessary to ascertain the weight of the precipitate produced by it, in order to learn the absolute proportion of tannin present in any specimen.

He next directed his attention to the discovery of some method by which the *gallic acid* might be separated from *extractive matter*, in cases where they exist in combination, but the enquiry was not successful; for, as he observes, it is difficult to render the *extractive* insoluble, so as to separate it, without at the same time decomposing the gallic acid. It is true that æther will dissolve the latter, without exerting much action upon the former; but then, he adds, whenever the gallic acid is in large quantities, this method will fail, "in consequence of that *affinity* which is connected with mass." Here then he adopts that celebrated theory of Berthollet,* which he afterwards so vigorously and successfully attacked.†

* Récherches sur les Lois de l'Affinité.—Mém. de l'Institut National, Tome III. p. 5.

† The masterly manner in which he combated the successive arguments of Berthollet upon this question is admirable. In the first place, he attacked the theory upon general principles, and then exposed the fallacy of the several experiments adduced in its

As general tests of the respective quantities of these two principles (gallic acid and extractive matter), he employed the solutions of the salts of alumina and those of the peroxidated salts of iron. The former of these precipitates *extractive*, without materially acting upon *gallic acid*, which is thrown down by the latter: the greatest care, however, must be taken not to add the iron in excess, as in that case the black precipitate formed will be re-dissolved, and an olive-coloured and clear fluid be only obtained.

He details the results of a number of experiments made upon galls, and ascertains the relative proportions of their several elements; and he proves that tannin may exist in such a state of combination in different substances as to elude the test of gelatine; in which case, to detect its presence, it is necessary to have recourse to the action of the diluted acids.

support. "Were the proposition correct, that *in all cases of decomposition in which two bodies act upon a third, that third is divided between them in proportion to their relative affinities, and their quantities of matter*, it is quite evident that there could be scarcely any definite proportions: a salt crystallizing in a strong alkaline solution would be strongly alkaline; in a weak one, less alkaline; and in an acid solution it would be acid." With regard to glasses and alloys, adduced by M. Berthollet as compounds of indefinite proportions, Davy answers—"It is not easy to prove, in such cases, that the elements are chemically combined, for the points of fusion of alkali, glass, and certain metallic oxides, are so near to each other, that transparent mixtures of them may be formed." The experiment upon which M. Berthollet laid great stress, viz. that a large quantity of potash will separate a small quantity of sulphuric acid from sulphate of baryta, Davy invalidates in a most complete manner. He says—"This experiment was made in contact with the atmosphere, in which carbonic acid is always present; and carbonate of potash and sulphate of baryta mutually decompose each other."

Sir Joseph Banks had concluded, from the sensible properties of *catechu*,* or *terra Japonica*, that it was rich in tannin: Davy confirmed this opinion by experiment. The leather tanned by it appeared to possess the same qualities as when tanned in the usual manner; and although this substance contains a small portion of extractive matter, yet the increase of weight of the skin was rather less than when solutions of galls were used.

In examining different barks, he was not able to procure from them any free gallic acid, but their infusions gave, on evaporation, tannin and extractive. The greater number of his experiments were made on the barks of the oak, the Leicester willow, the Spanish chestnut, the elm, and the common willow. The largest quantity of tannin he found to be contained in the interior, or white cortical layers; and the largest quantity of extractive matter in the exterior layers; the epidermis, or rough outward bark, did not contain either the one or the other.

From his general observations he is inclined to suppose that, in all the astringent vegetables, the tannin is of the same species, and that all the differences attributed to it depend upon its state of combination with other principles.

In applying the results of his experiments to the theory of tanning, he considers the process as simply

* Catechu is an extract obtained from the wood of a species of the *Mimosa* that grows in India, by boiling and subsequent evaporation. It is of two kinds; one from Bengal, the other from Bombay. The former contains rather less, the latter rather more, than half its weight of tannin. The remainder in both cases is a peculiar extractive matter mixed with mucilage.—P.

depending upon the union of the tannin with the matter of the skin, in such a manner as to form with it an insoluble compound. Gallic acid, he feels assured, does not produce any notable effects in the process; but he thinks that the quality of the leather depends, in some degree, upon the quantity of extractive matter it may imbibe.

Skin, combined with extractive matter only, would be increased in weight, become coloured, and be extremely flexible, but it would not be insoluble in water; and were it combined with tannin alone, it would be heavier and less supple than when both these principles enter into the compound.

He examines with great acuteness and precision some of the more popular opinions entertained by tanners, and brings his science to bear very satisfactorily upon several of their processes.

The grand secret, on which the profit of the trade mainly depends, is to give the hides the greatest increase of weight in the least possible time. To effect this, various schemes have been proposed, many of which, from the ignorance of the operators, instead of promoting, have defeated the object. Different *menstrua* have been suggested for expediting the process, and amongst them lime-water and the solutions of pearl-ash; but, as he has clearly shown, these two substances form compounds with tannin which are not decomposable by gelatine; whence it follows that their effects must be pernicious; and there is very little reason to suppose that any bodies will be found which, at the same time that they increase the solubility of tannin in water, will not likewise diminish its attraction for skin.

His experiments having proved that the saturated infusions of astringent barks contain much less extractive matter, in proportion to their tannin, than those which are weaker, it follows, that by quickly tanning the skin, we render the leather less durable. These observations show that there is some foundation for the vulgar opinion of workmen, concerning what is technically called the *feeding* of leather in the slow method of tanning.

Such is an outline of this interesting paper, in which the author has displayed the talent so characteristic of his mind—that of bringing science and art into useful alliance with each other. It forms, at this day, the guide of the tanner; and those who previously carried on the process by a routine of operations, of which they knew not the reasons, are now capable of modifying it, without the risk of spoiling the result. Many of those expedients which have been brought forward as novelties in later years, may be found in this paper; or, at least, have arisen out of the principles disclosed during his investigations.

It has been stated that, shortly after Davy's successful *début* as a lecturer, his manners underwent a change, and that, to the regret of his friends, he lost much of his native simplicity. On the 5th of February 1802, he had dined with Sir Harry Englefield at his house at Blackheath; and eighteen years afterwards, the worthy Baronet alluded to his interesting demeanour upon that occasion, in terms sufficiently expressive of his feelings—"It was the last flash of expiring Nature." It was natural that his best friends, on perceiving this change of manner, should entertain some apprehensions as to the deeper qualities of his heart. Mr. Purkis has placed in my hands the following letter addressed to him by Mr. Coleridge; it will interest the reader by the force and truth with which its talented writer characterises the perils which beset the elevated path of the young philosopher at the commencement of his career.

TO SAMUEL PURKIS, ESQ.

MY DEAR PURKIS, Nether Stowey, Feb. 17, 1803.

I RECEIVED your parcel last night, by post, from Gunville, whither (crossly enough) I am going with our friend Poole to-morrow morning. I do from my very heart thank you for your prompt and friendly exertion, and for your truly interesting letter. I shall write to Wedgwood by this post; he is still at Cote, near Bristol; but I shall take the *Bang* back with me to Gunville, as Wedgwood will assuredly be there in the course of ten days. Jos. Wedgwood is named the Sheriff of the County. When I have heard from Wedgwood, or when he has tried this *Nepenthe*, I will write to you. I have been here nearly a fortnight; and in better health than usual. Tranquillity, warm rooms, and a dear old friend, are specifics for my complaints. Poole is indeed a very, very good man. I like even his incorrigibility in small faults and deficiencies: it looks like a wise determination of Nature to let well alone; and is a consequence, a necessary one perhaps, of his immutability in his important good qualities. His journal, with his own comments, has proved not only entertaining but highly instructive to me.

I rejoice in Davy's progress. There are three Suns recorded in Scripture:—Joshua's, that stood still; Hezekiah's, that went backward; and David's, that went forth and hastened on his course, like a bridegroom from his chamber. May our friend's prove the latter! It is a melancholy thing to see a man, like the Sun in the close of the Lapland summer, meridional in his horizon; or like wheat in a rainy season, that shoots up well in the stalk, but does not *kern*. As I have hoped, and do hope, more proudly of Davy than of any other man; and as he has been endeared to me more than any other man, by the being a Thing of Hope to me (more, far more than myself to my own self in my most genial moments,)—so of course my disappointment would be proportionally severe. It were falsehood, if I said that I think his present situation most calculated, of all others, to foster either his genius, or the clearness and incorruptness of his opinions and moral feelings. I see two Serpents at the cradle of his genius, Dissipation with a perpetual increase of acquaintances, and the constant presence of Inferiors and Devotees, with that too great facility of attaining admiration, which degrades Ambition into Vanity—but the Hercules will strangle both the reptile monsters. I have thought it possible to exert talents with perseverance, and to attain true greatness wholly pure, even from the impulses of ambition; but on this subject Davy and I always differed.

When you used the word "gigantic," you meant, no doubt, to give me a specimen of the irony I must expect from my Philo-Lockian critics. I trust, that I shall steer clear of almost all offence. My book is not, strictly speaking, metaphysical, but historical. It perhaps will merit the title of a History of Metaphysics in England from Lord Bacon to Mr. Hume, inclusive. I confine myself to facts in every part of the work, excepting that which treats of Mr. Hume:—*him* I have assuredly besprinkled copiously from the fountains of Bitterness and Contempt. As to this, and the other works which you have mentioned, "have patience, Lord! and I will pay thee all!"

Mr. T. Wedgwood goes to Italy in the first days of May. Whether I accompany him is uncertain. He is apprehensive that my health may incapacitate me. If I do not go with him, (and I shall be certain, one way or the other, in a few weeks,) I shall go by myself, in the first week of April, if possible.

Poole's kindest remembrances I send you on my own hazard; for he is busy below, and I must fold up my letter. Whether I remain in England or am abroad, I will occasionally write you; and am ever, my dear Purkis, with affectionate esteem,

Your's sincerely,

S. T. COLERIDGE.

Remember me kindly to Mrs. Purkis and your children. T. Wedgwood's disease is not painful: it is a complete *tædium vitæ;* nothing pleases long, and novelty itself begins to cease to act like novelty. Life and all its forms move, in his diseased moments, like shadows before him, cold, colourless, and unsubstantial.

From the tone of the following letter, it may be presumed also, that Mr. Poole, to whom it is addressed, had expressed some anxiety upon the dangers to which his flattering station exposed him.

TO THOMAS POOLE, ESQ.

MY DEAR POOLE, London, May 1, 1803.

HAVE you no thoughts of coming to London? I have always recollected the short periods that you have spent in town, with a kind of mixed feeling of pleasure and regret.

In the bustling activity occasioned in cities by the action and re-action of diversified talents, occupations, and passions, our existence is, as it were, broken into fragments, and with you I have always wished for unbroken intercourse and continuous feeling.

* * * *

Be not alarmed, my dear friend, as to the effect of worldly society on my mind. The age of danger has passed away. There are in the intellectual being of all men, permanent elements, certain habits and passions that cannot change. I am a lover of Nature, with an ungratified imagination. I shall continue to search for untasted charms,—for hidden beauties.

My *real*, my *waking* existence is amongst the objects of scientific research: common amusements and enjoyments are necessary to me only as dreams, to interrupt the flow of thoughts too nearly analogous to enlighten and to vivify.

Coleridge has left London for Keswick; during his stay in town, I saw him seldomer than usual;

when I did see him, it was generally in the midst of large companies, where he is the image of power and activity. His eloquence is unimpaired; perhaps it is softer and stronger. His will is probably less than ever commensurate with his ability. Brilliant images of greatness float upon his mind: like the images of the morning clouds upon the waters, their forms are changed by the motion of the waves, they are agitated by every breeze, and modified by every sunbeam. He talked in the course of one hour, of beginning three works, and he recited the poem of Christobel unfinished, and as I had before heard it. What talent does he not waste in forming visions, sublime, but unconnected with the real world! I have looked to his efforts, as to the efforts of a creating being; but as yet, he has not even laid the foundation for the new world of intellectual forms.

When my Agricultural Lectures are finished, I propose to visit Paris, and perhaps Geneva. How I regret that circumstances had not enabled us to make the same tour at the same time! I think, at all events, I shall see you before the Autumn, on your own lands, amidst your own images and creations. Your affectionate friend,

HUMPHRY DAVY.

TO THE SAME.

MY DEAR POOLE, Royal Institution.

OFTEN, very often, in the midst of the tumults of the multitude in this great city, has my spirit turned in quietness and solitude towards you.

I hope soon to see you in Somersetshire, where

we may worship Nature, and the spirit that dwells in Nature, in your green fields and under your tranquil sky. My communications with you and Coleridge and Southey, and other ornaments of the great existing Being, have excited feelings which cheer me in the apathy of London, and which make me love human nature.

* * * *

Your account of the young man who murdered his wife, I read with deep interest. It is from such narratives of the conduct of common persons, that the laws of simple human nature must be deduced. Beings acted on by few objects, awakening in them few but deep passions, are the beings which Metaphysicians and Moralists ought to study; not those who exist in general life, having their energies and feelings so attached to multiplied and indefinite things—so mixed up and connected with myriads of circumstances, as to be imperceptible, unless by a microscopic moral eye. I am, &c. &c.

H. Davy.

From the former letter, we learn that Davy, at this period, proposed delivering some lectures on the Chemistry of Agriculture. From the memorandums of my late friend Mr. Arthur Young, the celebrated Secretary to the Board of Agriculture, I have succeeded, through the kindness of his daughter, in procuring the following extracts; the only source from which I have been able to obtain any correct information upon this point in his scientific life.—"May 15th, 1803. *Mem.* Two lectures by Mr. Davy have taken place, and been very well attended; they intend retaining him by a salary of a hundred pounds a year,—a very good plan."

Amongst the pamphlets at Bradfield Hall is a small quarto of fourteen pages, entitled, "Outlines of a Course of Lectures on the Chemistry of Agriculture, to be delivered before the Board of Agriculture, 1803." It was evidently only printed for private circulation amongst the members. At the same time, he printed a small pamphlet, containing an explanation of the terms used in chemistry, for the instruction of those amongst his audience who had not particularly directed their attention to the science.

The first lecture was delivered on Tuesday, May the 10th, at twelve at noon, and five others on the succeeding Tuesdays and Fridays.

In an address to the Board of Agriculture by Sir John Sinclair, delivered in April 1806, in reviewing the various objects to which the attention of the Board had been directed, he thus alludes to the subject:—"In the year 1802, when my Lord Carrington was in the chair, the Board resolved to direct the attention of a celebrated lecturer, Mr. Davy, to agricultural subjects; and in the following year, during the Presidency of Lord Sheffield, he first delivered to the members of this Institution, a course of lectures on the CHEMISTRY OF AGRICULTURE. The plan has succeeded to the extent which might have been expected from the abilities of the gentleman engaged to carry it into effect. The lectures have hitherto been exclusively addressed to the Members of the Board; but to such a degree of perfection have they arrived, that it is well worthy of consideration, whether they ought not to be given to a larger audience. If such an idea met with the approbation of the Board, a hall might be procured for that purpose, or a special course of lectures read in this room exclusively for strangers."

Davy would appear to have been very early impressed with the importance of Chemistry, in its various applications to Agriculture. Allusions are constantly made to it in his letters; and at the conclusion of his "RESEARCHES," he glances at this department of the chemistry of vegetation, and observes that, "although it is immediately connected with the art upon which we depend for subsistence, it has been but little investigated."

In his introductory lecture of 1802, he speaks more forcibly upon the subject.

"Agriculture, to which we owe our means of subsistence, is an art intimately connected with chemical science; for although the common soil of the earth will produce vegetable food, yet it can only be made to produce it in the greatest quantity, and of the best quality, by methods of cultivation dependent upon scientific principles.

"The knowledge of the composition of soils, of the food of vegetables, of the modes in which their products must be treated, so as to become fit for the nourishment of animals, is essential to the cultivator of land; and his exertions are profitable and useful to society, in proportion as he is more of a chemical professor. Since indeed the truth has been understood, and since the importance of agriculture has been generally felt, the character of the agriculturist has become more dignified, and more refined;—no longer a mere machine of labour, he has learned to think, and to reason. He is aware of his usefulness to his fellow-men, and he is become, at once, the friend of nature, and the friend of society."

His appointment, as chemical professor to the Board of Agriculture, was accompanied with the obligation of reading lectures before its members; which he continued to deliver every successive season for ten years, modifying and extending their views, from time to time, in such a manner as the progress of chemical discovery might render necessary.

These discourses were collated, and published in the year 1813, at the request of the President and members of the Board, and they form the only systematic work we, at present, possess on the subject. Its views, however, are too generally interesting to be briefly dismissed; I shall therefore enter more fully into their merits in a more advanced part of these memoirs.

His connexion with the Board necessarily brought him in contact with the practical agriculturists and capitalists of the day, with many of whom he formed friendships which lasted through life. With Mr. Coke of Holkham he became well acquainted, and generally formed one of the party at his annual sheep-shearing.* He was also a frequent visitor at

* In the 40th volume of the "Annals of Agriculture," an account is given of the Holkham Sheep-shearing for 1803, and in the list of the company is the name of "Mr. Professor Davy."—At the meeting of 1808, he was also present, and is mentioned as the great chemist, whose discoveries will immortalize his name.

Mr.

Woburn, and received from the Duke the means by which he was enabled to submit to the test of practice various theories which his science had suggested.

In a letter to Mr. Gilbert, dated October 1803, he says: "I have just quitted the coast of Sussex, where I have spent the last three weeks with Lord Sheffield, the worthy biographer of Gibbon." In fact, there was not a nobleman, distinguished for intellectual superiority, who did not feel a pride in receiving him as a guest; and he passed his vacations in the society of those exalted persons who, in possessing rank, fortune, and talents, felt that they only held such gifts from Providence, in trust for the welfare of their fellow-countrymen.

We can scarcely picture to ourselves a being

Mr. Coke, in the course of his speech after dinner, alluding to the question of long and short dung, said, "It is the opinion of a friend of mine, who sits near me, Professor Davy, and upon whose judgment, on account of his extensive chemical as well as other scientific knowledge, I place the highest reliance, that the manure carried immediately on the field, without being disturbed, will have a greater effect in exciting rapid vegetation, and in encouraging the growth of the turnip plant, than when applied in the ordinary manner; for, under such circumstances, it will not only be more moist and alkaline, but it will be protected from a loss of substance, amounting very nearly to one-third of its original bulk." Davy afterwards, in company with the Duke of Bedford, Lord William Russell, Lord Thanet, Sir Joseph Banks, and other agriculturists, inspected several farms.—In 1812, his health was drunk at the Woburn Sheep-shearing by the Duke of Bedford; and in the following festival it was proposed by Lord Hardwicke.

In the print of the "Woburn Sheep-shearing," published by Garrard, in 1811, No. 75 represents Davy; he is standing, in a listening attitude, behind Mr. Coke, who is conversing with Sir Joseph Banks, Sir John Sinclair, and Mr. Arthur Young.

upon whom fortune ever showered more favours than upon Davy, during this golden period of his career. Independent in an honourable competence, the product of his genius and industry; resident in the centre of all scientific information and intelligence; every avenue of knowledge, and every mode of observation open to his unwearied intellect, he must have experienced a satisfaction which few philosophers have ever before felt,—the power of pursuing experimental research to any extent, and of commanding the immediate possession of all the means it might require, without the least regard either to cost or labour. What a contrast does this picture afford to that which has been too faithfully represented as the more usual fate of the philosopher and man of letters, and which exhibits little more than the unavailing struggles of genius against penury! Instead of a life consumed in fruitless expectation of patronage and reward, we behold Davy, in the full bloom of reputation, courted by all whom rank, talent, or station, had rendered conspicuous.

His life flowed on like a pure stream, under a sky of perpetual sunshine,—not a gust ruffled its surface, not a cloud obscured its brightness. In the morning, he was the sage interpreter of Nature's laws; in the evening, he sparkled in the galaxy of fashion; and not the least extraordinary point in the character of this great man, was the facility with which he could cast aside the cares of study, and enter into the trifling amusements of society.—"*Ne otium quidem otiosum,*" was the exclamation of Cicero; and it will generally apply to the leisure of men actively engaged in the pursuits of science; but

Davy, in closing the door of his laboratory, opened the temple of pleasure. When not otherwise engaged, his custom was to play at billiards, frequent the theatre, or read the last new novel. In ordinary cases, the genius of evening dissipation is an arrant Penelope; but Davy, on returning to his morning labours, never found that the thread had been unspun during the interruption.

The following anecdote is well calculated to illustrate that versatility of talent of which I have frequently spoken, as well as the power he possessed of abstracting himself, without detriment, from the most elaborate investigations. A friend of the late Mr. Tobin called upon him at the Institution, and found him deeply engaged in the laboratory; their conversation turned upon "The Honey Moon," which was to be brought out on the following evening.* No sooner had Davy heard that, although pressing applications had been made to several of the poets of the day, a Prologue had not yet been written, than he instantly quitted the laboratory, and in two hours produced that which was recited on the occasion by Mr. Bartley, and printed in the first edition of the comedy. I insert it in this place.

No uniformity in life is found:—
In ev'ry scene varieties abound;
And inconsistency still marks the plan
Of that immortal noble being,—Man.
As changeful as the April's morning skies,
His feeling and his sentiments arise;

* "The Honey Moon" was produced at Drury Lane, on Thursday, the 30th of January, 1805.

And Nature to his wond'rous frame has given
The mingled elements of Earth and Heaven.
In diff'rent climes and ages, still we find
The same events for diff'rent ends design'd:
And the same passion diff'rent minds can move
To thoughts of sadness or to acts of love.
Hence Genius draws his novel copious store;
And hence the new creations we adore:
And hence the scenic art's undying skill
Submits our feeling to its potent will;
From common accidents and common themes
Awakens rapture and poetic dreams;
And, in the trodden path of life, pursues
Some object clothed in Fancy's loveliest hues—
To strengthen nature, or to chasten art,
To mend the manners or exalt the heart.
So thought the man whom you must judge to-night;
And as he thought, he boldly dared to write.
Not new the subject of his first-born rhyme;
But one adorn'd by bards of elder time;—
Bards with the grandest sentiments inspired—
Bards that in rapture he has still admired;
And tried to imitate, with ardour warm,
And catch the spirit of their pow'rful charm.
With loftiest zeal and anxious hope, he sought
To bring to modern times their strength of thought;
And, in their glowing colours, to display
The follies and the virtues of the day.
Whether his talents have his wish belied,
Your judgment and your candour must decide.
He, though your loftiest plaudits you should raise—
He cannot thank you for the meed of praise.
Rapture he cannot feel, nor fear, nor shame;
Connected with his love of earthly fame,
He is no more.—Yet may his memory live
In all the bloom that early worth can give!
Should you applaud, 'twould check the flowing tear
Of those to whom his name and hopes are dear.

But should you an unfinish'd structure find,
As in its first and rudest forms design'd,
As yet not perfect from the glowing mind,
Then with a gentle voice your censure spread,
And spare the living—spare the sacred DEAD!

Davy would appear to have frequently amused himself with writing sonnets, and inclosing them in letters to his several friends: the following letter will also show that he was ambitious of being considered a poet.

TO SAMUEL PURKIS, ESQ.

MY DEAR PURKIS,

I INCLOSE the little poem,* on which your praise has stamped a higher value, I fear, than it deserves.

If I thought that people in general would think as favourably of my poetical productions, I would write more verses, and would write them with more care; but I fear you are partial: I am very glad, however, that you like the little song; at some future period I will send you another.

With kind remembrances, unalterably your sincere friend,

H. DAVY.

On examining the laboratory notes made at this period, many of which, however, are nearly illegible from blots of ink and stains of acid, it would appear that his researches into the composition of mineral bodies were most extensive, and that he obtained many new results, of which he does not seem to

* The subject was "*Julia's Eyes*."

have availed himself in any of his subsequent papers. To borrow a metaphor from his favourite amusement, he treated such results as small fry, which he returned to their native element to grow bigger, or to be again caught by some less aspiring brother of the angle.

Had Davy, at this period of his life, been anxious to obtain wealth,* such was his chemical reputation, and such the value attached to his judgment, that, by lending his assistance to manufacturers and projectors, he might easily have realized it; but his aspirations were of a nobler kind — SCIENTIFIC GLORY was the grand object for which his heart panted: by stopping to collect the golden apples, he might have lost the race.

* I am assured by one of his earliest friends, that, at this period, he did not appear even to have an idea of the value of property. Any thing not immediately necessary to him he gave away, and never retained a book after he had read it.

CHAPTER V.

Sir Thomas Bernard allots Davy a piece of ground for Agricultural Experiments.—History of the Origin of the Royal Institution.—Its early labours.—Davy's Letters to Mr. Gilbert and to Mr. Poole.—Death of Mr. Gregory Watt.—Davy's passion for Fishing, with Anecdotes.—He makes a Tour in Ireland: his Letters on the subject.—His paper on the Analysis of the Wavellite.—His Memoirs on a new method of analysing Minerals which contain a fixed Alkali.—Reflections on the discovery of Galvanic Electricity.

VERY shortly after Davy had arrived in London, he formed an intimate friendship with Mr. (afterwards Sir Thomas) Bernard; and no sooner had he directed his attention to the subject of Philosophical Agriculture, than the worthy Baronet allotted him a considerable piece of ground near his villa at Roehampton, where, under his sole direction, numerous experiments were tried, many of which proved highly successful, and afterwards served for the illustration of various subjects in his work on AGRICULTURAL CHEMISTRY.

Although devoted as Davy was to the pursuits of science, he entered warmly into all political plans for improving the condition of the people, and advancing the progress of civilization. "No one," says his friend Mr. Poole, "was less a sectarian, if I may use the word, in religion, politics, or in science. He regarded with benevolence the sincere convictions of any class on the subject of belief, however they might differ from his own. In politics, he was the ardent friend of rational liberty. He gloried in the institutions of his country, and was anxious to see them maintained in their purity by timely and temperate reform." Indeed, in carefully analysing his mind, and tracing its developement, it appears that benevolence was one of its leading elements; the form in which it displayed its energies varying with the varying conditions of intelligence. In boyish life, his imagination, acting upon his zeal for the welfare of his species, delighted, as we have seen, in the ideas of encountering dragons, and quelling the might of giants; but as fancy paled with the light of advancing years, and the judgment presented distincter appearances, the philanthropic antipathy which had been directed to those chimeras of the nursery, was transferred to the two great oppressions of society, and in Superstition he saw the dragons—in Despotism, the giants that spread mischief and misery through the world.

Some of his early manuscripts are still in existence; and I shall here introduce a passage from one which has been lately transmitted to me by a gentleman resident in Penzance. The most trifling record becomes interesting, when we can trace in it the germ of a particular opinion, or the first symptom of a quality which may afterwards have distinguished its possessor.

"Science is as yet in her infancy; but in her infancy she has done much for man. The discoveries hitherto so beneficial to mankind have been generally effected by the energies of individual minds:—what hopes may we not entertain of the rapid progress of the happiness of man when illumination shall become general—when the united powers of a number of scientific men shall be employed in discovery! Every thing seems to announce the rapid advance of this period of improvement. The time is approaching when despotism and superstition, those enormous chains that have so long enfettered mankind, shall be annihilated,—when liberated man shall display the mental energies for which he was created. At that period, nations shall know that it is their interest to cultivate science, and that the benevolent philosophy is never separated from the happiness of mankind."

In his published writings, we discover evidences of the same tendency; he suffers no opportunity to escape which can enable him to enforce his principle, and he extracts from the most common as well as from the least probable sources, comparisons and analogies for its illustration. The ingenuity with which this is accomplished often surprises and delights us; the effect upon the reader is frequently not unlike that occasioned by the flashes of wit, to which it surely must be closely allied, if wit be correctly defined by Johnson "a combination of dissimilar images, or the discovery of occult resemblances in things apparently unlike." Is not this opinion strikingly illustrated by the happy turn given to his observations "upon the process of obtaining nitrous oxide from nitre,"—when he says,

"Thus, if the hopes which these experiments induce us to indulge do not prove fallacious, a substance which has heretofore been almost exclusively appropriated to the destruction of mankind, may become, in the hands of philosophy, the means of producing health and pleasure!"

Mr. Poole, who watched the whole of his progress from obscurity to distinction, and enjoyed his friendship for nearly thirty years, says, "To be useful to science and mankind was, to use his favourite expression, the pursuit in which he gloried. He was enthusiastically attached to science, and to men of science; and his heart yearned to promote their interests."

That Davy, with a mind so constituted, should have formed a strong and ardent attachment to Sir Thomas Bernard, and that this friendship should have been reciprocally cultivated, cannot be a matter of surprise.

I am happy in this opportunity of paying a tribute of respect to the memory of this most excellent person, with whom I had the pleasure of being well acquainted. His life was one continued scheme of active benevolence; and he merits a particular notice in these memoirs, as being one of the principal founders and patrons of the Royal Institution. Actuated by that noble and rational ambition which makes private pursuits subservient to public good, he directed all the energies of his mind, the influence of his station, and the resources of his wealth, towards promoting societies and schemes for encouraging the virtues and industry, and for ameliorating the condition, of the lower classes.

In the beginning of November 1796, in conjunction with the late Bishop of Durham, Mr. Wilberforce, and Mr. Elliot, he established the Society for bettering the condition of the Poor. As one of the primary objects of the original promoters of this society was the formation of an institution which might teach the application of science to the advancement of the arts of life, and to the increase of domestic comforts, a select committee was appointed from its body, in January 1799, for the purpose of conferring with Count Rumford on the means of carrying such a scheme into practical effect. This committee consisted of the Earl of Winchelsea, Mr. Wilberforce, Mr. Sullivan, the Bishop of Durham, Sir Thomas Bernard, and some other members of the society; and in a few weeks they completed the arrangements, circulated printed proposals, and collected the subscriptions, which gave birth to the Royal Institution of Great Britain, the future cradle of experimental science, and the destined scene of Davy's glory.

In addition to the general objects of promoting the arts and manufactures, and of advancing the taste and science of the country, its more immediate purpose was the improvement of the means of industry and domestic comfort among the poor.

That this benevolent design was constantly kept in sight may be shown by the several resolutions passed at the different meetings of the Managers, especially at that held in March 1800; when it was resolved to appoint *fourteen* different committees, for the purpose of scientific investigation and improvement; amongst which were the following:—

"For the investigation into the processes of making bread, and into the methods of improving it.

"For enquiring into the art of preparing cheap and nutritious soups for feeding the poor.

"For improving the construction of cottages, and cottage fire-places, and for improving kitchen fire-places, and kitchen utensils.

"For ascertaining, by experiment, the effects of the various processes of cookery upon the food of cattle.

"For improving the construction of lime-kilns, and the composition of mortar and cements," &c. &c.

So that the foundation and original arrangements of the Royal Institution were not only calculated to extend the boundaries of science, but to increase its applications, and to promote and improve those arts of life on which the subsistence of all, and the comfort and enjoyment of the great majority of mankind absolutely depend.

At this early period of its history, the Royal Institution presented a scene of the most animated bustle and exhilarating activity. Persons most distinguished in the various departments of science and art were to be seen zealously and liberally co-operating for the promotion and diffusion of public happiness, under the cheering beams of popular favour and exalted patronage. It was like 'a busy ant-hill in a calm sunshine.'

I shall only add, that Sir Thomas Bernard was the original promoter of the "School for the Indigent Blind;" of an institution for the protection and instruction of "Climbing Boys;" of a society for the relief of "Poor Neighbours in Distress;" of the "Cancer Institution;" and of the "London Fever Hospital."

The philanthropic Baronet was, moreover, the founder of the "British Institution," for promoting the Fine Arts in the United Kingdom; and he was also the originator of the "Alfred Club."

The vast range and practical utility of these exertions were duly appreciated by his contemporaries, who were ever ready to promote any scheme which had received the sanction of his patronage. It is an anecdote worthy of being preserved, that the late Sir Robert Peel called upon him one morning, and after a general conversation on the different philanthropic objects they had in view, said on leaving the room, he had to request that Sir Thomas would dispose of something for him, in any manner he thought most serviceable, and laid on the table an enclosure. After he had left the house, Sir Thomas was greatly surprised, on opening it, to find a bank-note of a thousand pounds.

The active zeal of Sir T. Bernard, like every other circumstance which exceeds the ordinary standard of our conduct, or becomes prominent from the rarity of its occurrence, called forth the wit as well as the admiration of his contemporaries. One of those modern travellers who delight in astonishing their auditors by incredible tales and marvellous anecdotes, happening to be in company with a noble lord as much distinguished for the playfulness of his wit as for the profundity of his learning, told the following improbable story: that, in a sequestered part of Italy, when pressed by hunger and fatigue, he sought refreshment and repose in a wild dwelling in the mountains, and was agreeably surprised at being offered a pie; but, horror of horrors!

on examining its contents he found — a human finger! — "Nothing more probable, Sir," interrupted his Lordship; "and I well know the person to whom that finger belonged—to Sir Thomas Bernard, Sir, for he had a finger in every pie."

The following letters will be read in this place with interest.

TO DAVIES GIDDY, ESQ.

MY DEAR SIR,

I AM now on my way to Christchurch, in company with Mr. Bernard, who was the founder, and has been the great supporter, of the Society for bettering the condition of the Poor.

In a conversation that has just passed between us, I mentioned the state of improvement of the Downs between Helston and Marazion, in consequence of grants of small portions of land to miners and other tenants for cultivation, many of which have, I believe, been made by Lord Dunstanville. Mr. Bernard expressed a desire to know what the effect of this plan had been on the condition of the persons thus raised into "property-men."

He is accumulating facts as to the manner in which the poor have been most effectually benefited, and to assist his labour would be to assist a good and most important cause; perhaps, you will have the goodness to give me a statement on this subject, which of course shall be used as you may think proper. You may likewise have similar facts nearer home, on your own estates.

I am convinced that the effects of enabling the

common labourer to acquire property must be striking, and must often have been an object of your contemplation.

In making any statement of these facts, you will probably think it right to mention some particular cases, with dates, names, and accounts of the quantities of lands, the nature of the improvements, &c.

In the reports of the "Society for bettering the condition of the Poor," there is one made on this minute plan of Lord Winchelsea's grants of land to cottagers, which conveys very full and useful information.

I trust to your kindness, and believe me

Your obliged,

H. DAVY.

The following letter was written by Davy after his return from an excursion to that beautiful district, the north-west of the county of Somerset.

TO THOMAS POOLE, ESQ.

MY DEAR POOLE, October, 1804.

I RETURNED to town a little while ago, not sorry to see the great city of activity and life; not sorry to see it, though I had just spent two months in enjoying a scenery beautiful and, to me, new; in witnessing much hospitality and unadulterated manners, and in gaining much useful information.

Mr. Bernard is writing a history of the poor. I have lived much with him at Roehampton since my return, and he has read to me part of his work, which is popularly eloquent, very intelligent, and full of striking and important truths; but pray say nothing of this, for it is likely that it will appear without his name: the facts will be strong, and perhaps to some people offensive.

I have received a letter from Coleridge within the last three weeks: he writes from Malta, in good spirits, and, as usual, from the depth of his being. God bless him!—He was intended for a great man; I hope and trust he will, at some period, appear as such.

I am working very hard at this moment, and I hope soon to send you some of the fruits of my labours. I am likewise devising some plans at our Institute, for the improvement of "this generation of vipers;" but, although I am so vain as to announce them, I will not be so tedious as to detail them.

In your answer, which I hope I shall soon receive, pray give me an account of the situation of "Poole's Marsh," with regard to the *Parrot*,* for I have mentioned the soil in a paper to the Board of Agriculture, which is now in the press.

I am, my dear Poole,
Your truly affectionate friend,
H. Davy.

In this year, Davy was deprived of one of his earliest and most attached friends, after a lingering illness, during which his symptoms, by the alternations which characterise consumption, had inspired his friends with hope, only to chill them with de-

* He alludes to a rich piece of land near the river Parrot; a specimen of the soil of which Mr. Poole had sent him for analysis.

spondency;—Gregory Watt terminated his earthly career.*

On the first impression which this melancholy event produced upon his feelings, Davy wrote a letter to his friend Clayfield, from which the following is an extract.

"I scarcely dare to write upon the subject—I would fain do what Hamlet does, when, in awe and horror at the ghost of his father, he attempts to call up the ludicrous feeling, but being unable to do so, he merely employs the words which are connected with it.—I would be gay, or I would write gaily, in alluding to the loss we have both sustained, but I feel that it is impossible. Poor Watt!—He ought not to have died. I could not persuade myself that he would die; and until the very moment when I was assured of his fate, I would not believe he was in any danger.

"His letters to me, only three or four months ago, were full of spirit, and spoke not of any infirmity of body, but of an increased strength of mind. Why is this in the order of Nature, that there is such a difference in the duration and destruction of her works? If the mere stone decays, it is to produce a soil which is capable of nourishing the moss and the lichen; when the moss and the lichen die and decompose, they produce a mould which becomes

* Gregory Watt was one of those philosophers to whose memory justice has not awarded its due. He was a meteor, whose light no sooner flashed upon us than it expired. His paper upon the gradual refrigeration of Basalt, alone entitled him to a distinguished rank amongst experimentalists. It was read before the Royal Society in May; and he expired in the following October.

the bed of life to grass, and to a more exalted species of vegetables. Vegetables are the food of animals, —the less perfect animals of the more perfect; but in man, the faculties and intellect are perfected,—he rises, exists for a little while in disease and misery, and then would seem to disappear, without an end, and without producing any effect.

"We are deceived, my dear Clayfield, if we suppose that the human being who has formed himself for action, but who has been unable to act, is lost in the mass of being: there is some arrangement of things which we can never comprehend, but in which his faculties will be applied.

"The caterpillar, in being converted into an inert scaly mass, does not appear to be fitting itself for an inhabitant of air, and can have no consciousness of the brilliancy of its future being. We are masters of the earth, but perhaps we are the slaves of some great and unknown beings. The fly that we crush with our finger, or feed with our viands, has no knowledge of man, and no consciousness of his superiority. We suppose that we are acquainted with matter, and with all its elements, and yet we cannot even guess at the cause of electricity, or explain the laws of the formation of the stones which fall from meteors. There may be beings,—thinking beings, near us, surrounding us, which we do not perceive, which we can never imagine. We know very little; but, in my opinion, we know enough to hope for the immortality, the *individual immortality of the better part of man.*

"I have been led into all this speculation, which you may well think wild, in reflecting upon the fate of Gregory! my feeling has given erring wings to my mind. He was a noble fellow, and would have been a great man.—Oh! there was no reason for his dying—he ought not to have died.

"Blessings wait on you, my good fellow! Pray remember me to Tobin, and, if you read this letter to him, protest, the moment he begins to argue against the immortality of man!

"I came yesterday from the borders of Dorsetshire, where I have been since Monday, seduced to travel by a friend. I was within sixty miles of you, and saw divers fair trout-streams: let the fish beware of me,—I shall be at them on Monday."

I have included this latter sentence in my extract, as being highly characteristic of the writer. His passion for angling betrayed itself upon all occasions; and the sport was alike his relief in toil, and his solace in sorrow. To his conversation, as well as to his letters, we may aptly apply the words of the Augustan poet:—

"Desinit in piscem——formosa superne."

Whenever I had the honour of dining at his table, the conversation, however it might have commenced, invariably ended on fishing; and when a brother of the angle happened to be present, you had the pleasure of hearing all his encounters with the finny tribe—how he had lured them by his treachery, and vanquished them by his perseverance. He would occasionally strike into a most eloquent and impassioned strain upon some subject which warmed his fancy; such, for example, as the beauties of mountain scenery; but before you could fully enjoy the prospect which his imagination had pictured, down he carried you into some sparkling stream, or rapid current, to flounder for the next half hour with a hooked salmon!

I remember witnessing, upon one of these occasions, a very amusing scene, which may be related as illustrative of some peculiarities of his temper. I believe all those who have accompanied Davy in his fishing excursions, will allow that no sportsman was ever more ambitious to appear skilful and lucky. Nothing irritated him so much as to find that his companions had caught more fish than himself; and if, during conversation, a brother fisherman surpassed him in the relation of his success, he betrayed similar impatience.

There happened to be present, on the occasion to which I allude, a skilful angler, and an enterprising chemist. The latter commenced on some subject connected with his favourite science; but Davy, who, generally speaking, disliked to make it a subject of conversation, suddenly turned to the angler, and related what he considered a very surprising instance of his success: his sporting friend, however, mortified him by the relation of a still more marvellous anecdote; upon which Davy as quickly returned to the chemist, who, in turn, again sent him back to the angler: and thus did he appear to endure the unhappy fate of the *flying fish,* who no sooner escapes from an enemy in the regions of air, than he is pursued by one equally rapacious in the waters.—But to return to the thread of our history.

In referring to the records of the Institution, it appears that in January 1805, Davy greatly enriched the cabinets of the Institution by a present of minerals. The following are the Minutes of the Committee upon this occasion:

"January 21, 1805—Mr. Hatchett reported that, in pursuance of the request of the Managers, he had inspected the minerals presented to the Royal Institution by Mr. Davy, and that the aggregate value (including the duplicate specimens) appears to him to exceed one hundred guineas."

"January 28.—The Managers took into consideration Mr. Hatchett's report at the last meeting, and resolved that Mr. Davy is entitled to the thanks of the Managers for having added so valuable a present to the collection of minerals belonging to the Institution."

On the 4th of February, it was Resolved—"That Mr. Davy be appointed Director of the Laboratory, at a salary of one hundred pounds a-year; by which his annual income from the Institution was raised to four hundred pounds. At this period he delivered a series of lectures on Geology, or on the chemical history of the earth; to which we find an allusion in the following letter.

TO THOMAS POOLE, ESQ.

MY DEAR POOLE, February, 1805.

I AM very much obliged to you for your last kind letter, and I thank you most sincerely for the exertion of your friendship at Bath. I thank you with very warm feelings.

I hope you will soon come to town; that you will stay a long time; and that we shall be very much together.

I paid your subscription to Arthur Young for the Smithfield Club. Pray, at all times, command me to do any thing I can for you in London:—you cannot teaze me; and though I am a very idle fellow, yet I can always work if the stimulus be the desire of serving such a friend as yourself.

I am giving my course of lectures on Geology to very crowded audiences. I take a great interest in the subject; and I hope the information given will be useful.

There has been no news lately from Coleridge; the last accounts state that he was well in the autumn, and in Sicily. On that poetic ground, we may hope and trust that his genius will call forth some new creations, and that he may bring back to us some garlands of never-dying verse. I have written to urge him strongly to give a course of lectures on Poetry at the Royal Institution, where his feeling would strongly impress, and his eloquence greatly delight. I am, my dear Poole, most affectionately Yours,

H. Davy.

On the 20th of May, in this year, Mr. Hatchett reported to the Managers of the Institution—"that Mr. Davy proposed making a journey into Wales and Ireland this summer, having in view to collect specimens for enriching the mineralogical cabinets;" in consequence of which it was Resolved—"That the sum of one hundred pounds be entrusted to Mr. Davy to purchase minerals, and to defray the incidental charges; and that the boy of the Laboratory,

William Reeve,* be ordered to attend him on his tour, and that the steward be directed to defray his expenses.

From the following letters, it would appear that, having accomplished his purpose of visiting Ireland, he made a rapid journey into Cornwall for the sake of seeing his mother and sisters.

TO DAVIES GIDDY, ESQ.

Okehampton, September 1805.

MY DEAR SIR,

I AM accompanying my friend Mr. Bernard in a tour through the West of England, and I hope we shall reach Penzance in two or three days.

Mr. Bernard wishes much for the honour of your acquaintance, and I trust you will permit me to have the pleasure of making you known to him. Much kindness and long knowledge of him, may have made me partial to that gentleman, and may perhaps influence me when I say, that there is not a more patriotic, good, and public-spirited man in Great Britain.

I came from Ireland by the western road, about a fortnight ago. My expectations were fully satisfied with the appearances of the "Giant's Causeway." The arrangements of rocks of the Northern Cape of Ireland appear to me to present facts equally irreconcilable upon either the Plutonic or Neptunian theory; and I am convinced that general fanciful

* There are some circumstances of interest connected with the history of this young man. He possessed much chemical talent; but during his residence in Ireland he was converted to the Catholic religion, and is at this time a Catholic priest in some part of the Continent.

theories will lose ground in proportion as minute observations are multiplied.

The Irish are a noble race, degraded by slavery, and bearing the insignia of persecution, extreme savageness, or the lowest servility; yet they are ingenious and active, and seem to me to possess all the elements of power and usefulness; but amongst the lower orders there is a most unfortunate equality, destructive of all great and efficient exertion; and amongst the higher classes the greatest degree of activity is awakened only by the desire of imitating the English, and that not so much in their virtues and talents, as in their luxuries and follies.

I hear from all quarters of the good effects of your late exertions in Parliament. May your efforts tend to establish the reign of good sense and pure philosophy, in a place where they have been too often found to yield to empty sounds!

Yours, &c. H. DAVY.

TO THOMAS POOLE, ESQ.

MY DEAR POOLE, London, Oct. 9, 1805.

I MADE a very rapid journey to Cornwall with Mr. Bernard, merely for the sake of showing him the country, and for the purpose of spending a week with my mother and sisters.

We made an effort to come to you at Nether Stowey, but the people at Bridgewater would not take us round, through Stowey, to Taunton, without four horses; and at all events we could only have spent two or three hours with you; and it is difficult to say whether the pleasure of meeting, or the regret at parting so soon, would have been the greatest. I long very much for the intercourse of a week with you. I have very much to say about Ireland. It is an island which might be made a new and a great country. It now boasts a fertile soil, an ingenious and robust peasantry, and a rich aristocracy; but the bane of the nation is the equality of poverty amongst the lower orders. All are slaves without the probability of becoming free; they are in the state of equality which the *Sans-culottes* wished for in France; and until emulation and riches, and the love of clothes and neat houses, are introduced amongst them, there will be no permanent improvement.

Changes in political institutions can at first do little towards serving them. It must be by altering their habits, by diffusing manufactories, by destroying *middle-men*, by dividing farms,* and by promoting industry by making the pay proportioned to the work. But I ought not to attempt to say any thing on the subject when my limits are so narrow; I hope soon to converse with you about it.

I found much to interest me in geology in Ireland, and I have brought away a great deal of information, and many specimens.

I shall now be in London till Christmas, with the exception of next week, which I am obliged to pass in Bedfordshire. I am, my dear Poole,

Most affectionately your's,

H. DAVY.

* He means that the *middle-men* being discontinued, their large allotments should be divided into farms of convenient extent, the occupiers of which should rent immediately from the owners of the soil.

After the Giant's Causeway, the scenery which called forth Davy's greatest admiration in Ireland was that of Fair-Head. To an enthusiastic lover of the wild and sublime features of Nature, an object of greater interest could scarcely be presented than a vast promontory, the summit of which rises five hundred feet above the sea, and at whose base lies a waste of rude and gigantic columns, swept by the hand of Time from the mountain to which they formerly belonged.

The following fragment, written by Davy at the time, has been placed in my hands by Mr. Greenough.

"———— But chiefly thee, Fair-Head!
Unrivall'd in thy form and majesty!
For on thy loftiest summit I have walk'd
In the bright sunshine, while beneath thee roll'd
The clouds in purest splendour, hiding now
The Ocean and his islands—parting now
As if reluctantly: whilst full in view
The blue tide wildly roll'd, skirted with foam,
And bounded by the green and smiling land,
The dim pale mountains, and the purple sky.
Majestic cliff! thou birth of unknown Time,
Long had the billows beat thee, long the waves
Rush'd o'er thy hollow'd rocks, ere life adorn'd
Thy broken surface, ere the yellow moss
Had tinted thee, or the mild dews of Heaven
Clothed thee with verdure, or the eagles made
Thy cave their aëry: so in after time
Long shalt thou rest unalter'd mid the wreck
Of all the mightiness of human works;
For not the lightning nor the whirlwind's force,
Nor all the waves of ocean, shall prevail
Against thy giant strength—and thou shalt stand
Till the Almighty voice which bade thee rise
Shall bid thee fall."

Amongst Davy's letters to Mr. Gilbert, in the years 1804 and 1805, I find several upon the subject of the elastic force of steam, at different temperatures, with reference to Mr. Trevitheck's improvements in the steam-engine; in one of which he says, "I shall be extremely happy to hear of the results of your enquiries, and I hope you will not confine them to your friends, but make them public. Whenever speculative leads to practical discovery, it ought to be well remembered, and generally known: one of the most common arguments against the philosophical exercise of the understanding is, *Cui bono?* It is an absurd argument, and every fact against it ought to be carefully registered. Trevitheck's engine will not be forgotten; but it ought to be known and remembered that your reasonings and mathematical enquiries led to the discovery."

On the 28th of February 1805, was read before the Royal Society, and published in the Transactions of that year, a paper entitled, "An Account of some analytical Experiments on a mineral production from Devonshire, consisting principally of Alumina and Water; by Humphry Davy, &c."

This mineral was first discovered by Dr. Wavel, in small veins and cavities, in a tender argillaceous slate, near Barnstaple in Devonshire. At first it was considered as a species of *Zeolite*, until Mr. Hatchett concluded, from its geological position, that it did not belong to that family of minerals. Dr. Babington subsequently suspected from its physical characters, and from some of its habitudes with acids, that it was a mineral not before described, and accordingly placed a quantity of it in Davy's hands for analysis; who, on finding in its composition little more than clay and water, proposed to change the name of *Wavellite* for that of *Hydrargyllite*, as better expressive of its chemical nature. He however, at the same time, alludes to traces of an acid which he was unable to identify.

In a letter to Mr. Nicholson, dated Killarney, June 15, 1806, and which was afterwards published in his Journal, Davy refers to this fact in the following manner:—

DEAR SIR,

I SHALL feel much obliged to you to mention that I have found the acid which exists in minute quantities in Wavellite to be the *Fluoric acid*, in such a peculiar state of combination as not to be rendered sensible by sulphuric acid. I am, &c.

H. DAVY.

My late friend the Reverend William Gregor, having found the Wavellite at Stenna Gwynn, in Cornwall, submitted it to experiment, and the result certainly established the conclusion of the presence of fluoric acid, though not rendered apparent by the usual tests. The facts were transmitted to the Royal Society, and published in a paper entitled, "On a mineral Substance, formerly supposed to be Zeolite; by the Reverend William Gregor."

The subsequent experiments of Berzelius, however, cleared away the obscurity in which the subject was still involved. He showed that this mineral not only contained in its composition a small portion of the *neutral fluate of alumina*, but he demonstrated the presence of a *sub-phosphate* of that earth, to no inconsiderable an amount. Much has been said of the error committed on this occasion by Davy, in overlooking thirty-three per cent. of phosphoric acid; but the *phosphate of alumina* is a body that might very easily have escaped notice at a period when mineral analysis was in a far less advanced state than it is at present.

On the 16th of May 1805, Davy communicated to the Royal Society a paper "On the method of analyzing Stones containing a fixed Alkali, by means of the Boracic Acid." This method was founded upon two important facts: first, on the considerable attraction of boracic acid for the different simple earths at the heat of ignition; and, secondly, on the facility with which the compounds so formed are decomposed by the mineral acids. The processes are extremely simple, and the method must be considered as having advanced the art of mineral analysis.

For this and his preceding papers, the President and Council of the Royal Society adjudged to him their Copley medal.

In 1806, Mr. Poole, having consulted Davy on the subject of a Mine occurring near Nether Stowy, received from him the following letter, which is interesting from the political opinions it displays.

TO THOMAS POOLE, ESQ.

MY DEAR POOLE,

WHAT you have written concerning the indifference of men with regard to the interest of the species in future ages, is perfectly just and philosophical; but the greatest misfortune is, that men do not attend even to their own interest, and to the interest of their own age in public matters. They think in moments, instead of thinking, as they ought to do, in years; and they are guided by expediency rather than by reason. The true political maxim is, that the good of the whole community is the good of every individual; but how few statesmen have ever been guided by this principle! In almost all governments, the plan has been to sacrifice one part of the community to other parts:—sometimes, the people to the aristocracy; at other times, the aristocracy to the people;—sometimes, the Colonies to the Mother-country; and at other times, the Mother-country to the Colonies. A generous enlightened policy has never existed in Europe since the days of Alfred; and what has been called "the balance of power"—the support of civilization,—has been produced only by jealousy, envy, bitterness, contest, and eternal war, either carried on by pens or cannon, destroying men morally and physically! But if I proceed in vague political declamation, I shall have no room left for the main object of my letter—your Mine. I wish it had been in my power to write decidedly on the subject; but your county is a peculiar one: such indications would be highly favourable in Cornwall; but in a *shell-limestone* of late formation, there have as yet been no instances of great copper mines. I hope, however, that your mine will produce a rich store of *facts*.

Miners from Alston Moor, or from Derbyshire, would understand your country better than Cornish miners, for the Cornish shifts are wholly different from yours. It would be well for you to have some workmen at least from the North, as they are well acquainted with *shell-limestone.*

The Ecton copper mine in Staffordshire is in this rock: it would be right for you to get a plan and a history of that mine, which might possibly assist your views.

Had I been rich, I would adventure; but I am just going to embark with all the little money I have been able to save for a scientific expedition to Norway, Lapland, and Sweden. In all climes, I shall be your warm and sincere friend,

H. DAVY.

On the death of Dr. Edward Whitaker Gray, Secretary of the Royal Society, Davy was elected into that office, at an extraordinary meeting of the Society, on the 22nd of January 1807; and at the same time he was elected a member of the Council.

We are now advancing to that brilliant period in the history of our philosopher, at which he effected those grand discoveries in science, which will transmit his name to posterity, associated with those of Newton, Bacon, Locke, and the great master-spirits of every age and country:—I speak of his developement of the LAWS OF VOLTAIC ELECTRICITY.

I approach the subject with that diffidence which the contemplation of mighty achievements must ever produce in the mind of the historian, when he compares the extent and magnitude of his subject with the limited and feeble powers which are to describe them.

As the advantages afforded by the history of any

great discovery consist as much in exhibiting, step by step, the intellectual operations by which it was accomplished, as in detailing its nature and applications, or in examining its relations with previously established truths; so shall I be unable to preserve a chronological succession in the examination of those several memoirs which he presented to the Royal Society, without breaking asunder that fine intellectual thread, by which his mind was conducted through the intricate paths of nature from known to unknown phenomena. For this reason, although I announced, according to the date of its publication, the subject of his first paper on electricity, I deferred entering upon its examination, until I might be able to bring into one uninterrupted view the whole enquiry, in all its branches and bearings.

It is impossible to enter upon the subject of galvanism, or Voltaic electricity, without recurring to the circumstance which first betrayed the existence of such an energy in nature, and to the sanguine expectations which the discovery so naturally excited.

On witnessing the powerful contraction of a muscular fibre by the mere contact of certain metals, it was rational to conclude, that the nature and operation of the mysterious power of vital irritability might, at length, be discovered by a new train of scientific research. It is a curious fact, that an experiment so full of promise to the physiologist should have hitherto failed in affording him any assistance in his investigations; while the chemist, to whom it did not, at first, appear to offer any one single point of interest, has derived from it a new and highly important instrument of research, which has already, under the guidance of Davy, multiplied discoveries with such rapidity, and to such an extent, that it is not even possible to anticipate the limits of its power.

We have here, then, another striking instance of a great effect produced by means apparently insignificant. Who could have imagined it possible, that the spasmodic action occasioned in the limb of a frog, by the accidental contact of a pair of scissors, should have become the means of changing the whole theory of chemistry—of discovering substances, whose very existence was never suspected—of explaining the anomalous associations of mineral bodies in the veins of the earth—of protecting surfaces of metal from the corrosive action of the elements—of elucidating the theories of volcanoes and earthquakes—and, may we not add? of leading the way to a knowledge of the laws of terrestrial magnetism!

Such an unexpected extension of an apparently useless fact should dispose us to entertain a kinder regard for the labours of one another, and teach us to judge with diffidence of the abstract results of science. A discovery which may appear incapable of useful application to-day, may be our glory to-morrow,—it may even change the face of empires, and wield the destiny of nations.

The conic sections of Apollonius Pergæus remained useless for two thousand years: who could have supposed that, after the lapse of twenty centuries, they would have formed the basis of astronomy?—a science giving to navigation safety, guiding the pilot through unknown seas, and tracing for him in the heavens an unerring path to his native shores.

Some apology may be necessary for this digression; but, I confess, the subject has always appeared to me to be capable of much interesting illustration, and I heartily concur in the opinion expressed by the accomplished author of "Lettres à Sophie"— "*L'Histoire des grands effets par les petites causes ferait un livre bien curieux.*"

CHAPTER VI.

The History of Galvanism divided into six grand Epochs.—Davy extends the experiment of Nicholson and Carlisle.—His Pile of one metal and two fluids.—Dr. Wollaston advocates the doctrine of oxidation being the primary cause of Voltaic Phenomena. —Davy's modification of that theory.—His Bakerian Lecture of 1806.—He discovers the sources of the Acid and Alkaline matter eliminated from water by Voltaic action.—On the nature of Electrical decomposition and transfer.—On the relations between the Electrical energies of bodies, and their Chemical Affinities.—General developement of the Electro-chemical Laws.—Illustrations, Applications, and Conclusions.

THE History of Galvanism may be divided into six grand epochs; each being distinguished by the discovery of facts variously interesting from their novelty, and from the extent and importance of their applications.

It cannot be expected that I should enter into a minute history of the science; such a labour would require a distinct work for its accomplishment. I shall therefore follow the plan of the architect, who, in presenting a finished drawing of a part, sketches a faint outline of the whole edifice to which it belongs, in order that its fair proportions may appear in proper breadth and relief.

The FIRST EPOCH may be considered as arising out of the fundamental fact discovered by Galvani in 1790 — that the contact of two different metals with the nerve of a recently killed frog will excite distinct muscular contractions.

THE SECOND EPOCH may be dated from the discovery of what might be termed *Organic* Galvanism, or the production of its influence, without the presence of animal organs, by the peculiar action of metals upon water, as first observed by Dr. Ash.

THE THIRD EPOCH will long be celebrated on account of the discovery of the accumulation of the Galvanic power, by the invention of the pile of Volta, made known in the first year of the present century, and which so distinctly exhibited the analogy between Galvanism and Electricity, that the energy thus excited is now generally spoken of as "*Voltaic Electricity*."

THE FOURTH EPOCH may be considered as founded upon the knowledge of the general connexion between the excitement of Voltaic electricity and chemical changes.

THE FIFTH EPOCH is exclusively indebted for its origin to Davy—the establishment of the general law, that Galvanism decomposes all compound bodies, and that the decomposition takes place in a certain determinate manner.

THE SIXTH AND LAST EPOCH is founded upon the discovery of the relations subsisting between electricity and magnetism; giving origin to a new branch of science, which has been distinguished by the name of "ELECTRO-MAGNETISM."

Galvani,* from the moment of his first discovery,

* The simple fact relating to the action of metals on the animal organs was certainly not first observed by Galvani, but by

always referred the effects he produced to an electrical origin; but he considered that the metals employed merely acted as conductors, which effected a communication between the different parts of an animal, naturally, or by some process of nature, in opposite states of electricity, and that the muscular contractions took place during the restoration of the equilibrium.

Until the researches of Dr. Ash,* Ritter, Fabroni, and Creve, had been made known, the Galvanic influence was generally considered as existing only in the living organs of animals, from which it might be elicited by certain processes.

In the Bakerian Lecture† read before the Royal

Sulzer, who has described the sensation of taste produced by the contact of lead and silver with the tongue, in his *Théorie des Plaisirs*, in 1767.

* M. Humboldt (*Ueber die gereize Faser*, l. 473, 1797,) quotes part of a letter from Dr. Ash, in which it is said that, "if two finely polished plates of homogeneous zinc be moistened and laid together, little effect follows; but if zinc and silver be tried in the same way, the whole surface of the silver will be covered with oxidated zinc. Lead and quicksilver act as powerfully upon each other, and so do iron and copper. M. Humboldt says, that, in repeating this experiment, he saw air bubbles ascend, which he supposes to have been hydrogen gas from the decomposition of water.

† As this lecture will be frequently mentioned in the progress of these Memoirs, in connexion with most important discoveries, it may be interesting to the reader to learn something of its foundation and design. I have therefore collected the necessary information from the Minutes of the Royal Society. Mr. Baker is well known in the history of Science, as an accurate observer with the microscope, and as the author of several works on the subject. By his will, dated July 1763, he bequeathed the sum of one hundred pounds, the interest of which he directed "to be applied for an

Society in 1826, Davy, in giving a retrospective view of the progress of Electro-chemical Science, very justly remarks, that the true origin of all that has been done in this department of philosophy was the accidental discovery of Nicholson and Carlisle, of the decomposition of water by the pile of Volta, on the 30th of April, in the year 1800; which was immediately followed by that of the decomposition of certain metallic solutions, and by the observation of the separation of alkali on the negative plates of the apparatus. Mr. Cruickshank, in pursuing these experiments, obtained many new and important results, such as the decomposition of the *muriates of magnesia*, *soda*, and *ammonia*; and also observed the fact, that alkaline matter always appeared at the negative, and *acid* matter at the positive pole.*

No sooner had Davy become acquainted with the curious experiments of Nicholson and Carlisle, than,

Oration or Discourse, to be read or spoken yearly by some one of the Fellows of the Royal Society, on such parts of Natural History, or Experimental Philosophy, at such time, and in such manner, as the President and Council of the said Society shall please to order and appoint; on condition, nevertheless, that if any one year shall pass after the payment of the said hundred pounds, without such oration or discourse having been read or spoken at some Meeting of the said Royal Society, the said hundred pounds shall then become forfeited, and shall be repaid by the said Society to his executors," &c. Baker died in November 1774, and in the following year a Fellow was nominated to read the lecture. It is a whimsical circumstance, that the first lecturer should have been PETER WOULFE, the last of the alchemists. The names of the successive lecturers were as follow:—Dr. Ingenhouz, Mr. Cavallo, Mr. Vince, Dr. Wollaston, Dr. Young, Sir H. Davy, Mr. Brande, Captain Kater, Captain Edward Sabine, and Mr. Herschel.

* Nicholson's Journal, vol. iv. p. 190.

as we learn from his letter to Mr. Gilbert,* bearing the date of July 1800, he proceeded to repeat them. Indeed, it was the early habit of his mind not only to originate enquiries, but without delay to examine the novel results of other philosophers; and in numerous instances it would seem, that he only required to confirm their accuracy before he succeeded in rendering the application of them subservient to farther discovery. This was certainly the case with respect to the subject before us: he was a discoverer as soon as he became an enquirer. It is admirable to observe with what a quick perception he discovered the various bearings of a new fact, and with what ingenuity he appropriated it for the explanation of previously obscure phenomena. In referring to the "Additional Observations" appended to his "Chemical Researches," we shall find that the moment he became acquainted with the experiments of Dr. Ash, he proceeded to enquire how far the fact, previously noticed by himself, of the conversion of nitrous *gas* into nitrous *oxide*, by exposure to wetted zinc, might depend upon galvanic action.

In the month of September 1800, he published his first paper on the subject of Galvanic Electricity, in Nicholson's Journal, which was followed by six others, in which he so far extended the original experiment of Nicholson and Carlisle, as to show that oxygen and hydrogen might be evolved from separate portions of water, though vegetable and even animal substances intervened; and conceiving

* See page 85.

that all decompositions might be *polar*, he electrized different compounds at the different extremities, and found that sulphur and metallic bodies appeared at the *negative* pole, and oxygen and azote at the *positive* pole, though the bodies furnishing them were separated from each other. Here was the dawn of the Electro-chemical theory.

In a letter to Mr. Gilbert, already printed in these Memoirs,* he announced his opinion that Galvanism is a process principally chemical; and in a subsequent communication† to the same gentleman, written on the eve of his departure from Bristol to the Royal Institution, we discover a farther developement of the same theory, which, although modified by future researches, became, as we shall hereafter find, materially instrumental in establishing juster views of the nature of Voltaic action.

As soon as it was discovered that galvanic power might be excited by the contact of metals, without the interposition of animal organs, it was imagined that the electricity was set in motion by the contact of bodies possessing different conducting powers, without any reference to the chemical action which accompanied the process. This theory was naturally suggested by the fact discovered by Mr. Bennett several years before—that *electricity is excited by the mere contact of different metals:* thus, when a plate of copper and another of zinc, each furnished with an insulating glass handle, are made to touch by their flat surfaces, the zinc, after separation, exhibits *positive*, and the copper *negative* electricity.

* See page 110. † Page 118.

In this case, it is fair to conclude that a certain quantity of electricity had moved from the copper to the zinc.

On trying other metals, Volta found that similar phenomena arose; from which property such bodies have been denominated "*motors*" of electricity, and the process which takes place *electro-motion:* terms which have since been sanctioned and adopted by Davy.

It is on this transference of electricity from one surface to another, by simple contact, that Volta explains the action of the pile invented by himself, as well as that of all similar arrangements. The interposed fluids, on this hypothesis, have no effect as chemical agents, in producing the phenomena; they merely act as conductors of the electricity.

We have seen how early Davy had observed the intimate connexion subsisting between the electrical effect, and the chemical changes going on in the pile, and that he accordingly drew the conclusion of the dependence of the one upon the other. In fact, the most powerful Voltaic combinations are those formed by substances that act chemically upon each other with the greatest energy; while such as undergo no chemical change exhibit no electrical powers: thus zinc, copper, and nitric acid form a powerful battery; whilst silver, gold, and water, which do not act upon each other, produce no sensible effect in a series of the same number.

Although, in this obscure region of research, we are as yet unable to discover the nature of the power by which electricity is accumulated, it was a considerable step towards a true theory to have ascertained the insufficiency of the proposition that had been offered in explanation of the phenomena.

An investigation into the chemical activity of the pile led Davy to the discovery of a new series of facts, to which we find an allusion in his former letters to Mr. Gilbert, and which subsequently formed the basis of his first communication read before the Royal Society on the 18th of June in the same year.

All the combinations analogous to the Voltaic pile had hitherto consisted of a series containing, at least, *two* metallic bodies, (or one metal and charcoal,) and a stratum of fluid. Davy discovered that an accumulation of galvanic energy, exactly similar to that in the common pile, might be produced by the arrangement of *single* metallic plates with *different* strata of fluids; so that, instead of composing a battery with *two* metals and *one* fluid, he succeeded in constructing it with *one* metal and *two* fluids; provided always that oxidation, or some equivalent chemical change, should proceed on one of the metallic surfaces only.

In describing these combinations of a single metal with two fluids, he divides them into three classes, following in the arrangement the order of time with regard to their discovery.

In the First Class, one side of the metallic plate is oxidated; in the Second, a sulphuret is formed on one of its surfaces; and in the Third, both sides are acted upon, the metal becoming a *sulphuret* on one of its surfaces, and an *oxide* on the other.

The apparatus which he employed for these experiments is preserved in the laboratory of the Royal

Institution. It consists of a trough, containing grooves capable of receiving the edges of the different plates necessary for the arrangement, one half of which are composed of horn, the other half of some one metal.

When the apparatus was used, the cells were filled, in the galvanic order, with the different solutions, according to the class of the combination, and connected in pairs with each other by slips of moistened cloth carried over the non-conducting plates.

At the meeting of the Royal Society, following that on which the above interesting facts were communicated, Dr. Wollaston presented a memoir of considerable importance, entitled, "Experiments on the Chemical Production and agency of Electricity;" in which he strongly advocates the truth of that theory which recognises metallic oxidation as the *primary* cause of the Voltaic phenomena. This paper is also farther important as it proves, by most ingeniously devised experiments, not only the similarity of the means by which both common and galvanic electricity are excited, but also the resemblance existing between their effects; showing, in fact, that they are both essentially the same, and confirming the opinion, that all the apparent differences may depend upon differences in intensity and quantity.

Acting upon this principle, Dr. Wollaston succeeded in producing a very close imitation of the chemical action of galvanism by common electricity; such, for example, as the decomposition of water, and other effects of oxidation and deoxidation.* In

* M. Bonijol of Geneva has lately succeeded in effecting the decomposition of *Potash* and the *Chloride of Silver* by ordinary elec-

the prosecution of this train of research, he displays, in a very striking manner, that attention to minute arrangement which so remarkably characterised all his manipulations. I particularly allude to the expedients by which he reduced the extremity of a gold wire, in order to apportion the strength of the electric charge to the quantity of water submitted to its influence.

Although it is now very generally admitted, that the chemical agency of the fluids upon the metals employed is highly essential to the maintenance of Voltaic action, there still remains considerable doubt as to how far we are entitled to regard it as the first in the order of phenomena.

At a later period of his researches, Davy suggested as a correction, or rather modification, of the

tricity. His process consists in placing these substances in a very narrow glass tube, and in then passing a series of electric sparks from the ordinary machine through them. The electricity was conducted into the tube by means of two metallic wires fixed into the ends. When a quick succession of electric sparks had taken place for about five or ten minutes, the tube containing chloride of silver was found to contain reduced silver; and when potash had been submitted to the electric current, then the Potassium was seen to take fire as it was produced. The same philosopher has likewise contrived to decompose water by atmospheric electricity. The electricity, in this case, is collected from the atmosphere by means of a very fine point fixed at the extremity of an insulated rod; the latter is connected with the apparatus, in which the water is to be decomposed, by a metallic wire, of which the diameter does not exceed 1-50th of an inch. In this way the decomposition of the water proceeds in a continuous and rapid manner, although the atmospheric electricity be not strong. Stormy weather, it is said, is quite sufficient for the purpose.—*Bib. Univers.* 1830, p. 213. and *Royal Institution Journal*, No. 2.

theory of Volta, that the electro-motion produced by the contact of the metals might be the primary cause of the chemical changes; and that such changes were in no other way efficient, than in restoring the electric equilibrium thus disturbed: it was farther held, that this equilibrium could not be permanent, that it could in fact be only momentary; since, in consequence of the imperfect conducting power of the interposed fluid, the zinc and copper-plates, by their electro-motive power, would again assume their opposite states of electricity; and that these alternate changes would occur, as long as any of the fluid remained undecomposed. In a Voltaic arrangement, then, there would appear to exist, if the expression may be allowed, a kind of electrical see-saw; the apposition of the metals destroying the equilibrium, and the resulting chemical changes again restoring it. It has, however, been very justly observed, that the application of electricity, as an instrument of chemical decomposition, has most fortunately no connexion with such theories, and that the study of its effects may be carried on without reference to any hypothetical notions concerning the origin of the phenomena.

An interval of nearly five years had elapsed between the first communication which Davy made on this subject, and the Bakerian lecture which is immediately to be considered. During this period several new facts had been added by different experimentalists, but they were scattered, disjointed, and totally unconnected with each other by any rational analogies.

The constant appearance of acid and alkaline matter in pure water, when submitted to the influence of the Voltaic pile, gave rise to the most extravagant speculations and discordant hypotheses. Various statements were made, both in Italy and England, respecting the *generation* of muriatic acid, and that of the fixed alkalies, under these circumstances. Mr. Sylvester affirmed, that if two separate portions of water were electrised out of the contact of substances containing alkaline or acid matter, acid and alkali would, nevertheless, be produced.

Some philosophers sought to explain the phenomenon from the salts contained in the fluids of the trough, which they imagined might, by some unsuspected channel, find their way into the water under examination. Others believed that they were actually *generated* by the union of the electric fluid with the water, or with one or both of its elements; so that, up to the time of Davy's masterly researches, the subject was involved in the greatest obscurity; and whether the saline matter was liberated from unknown combinations, or at once formed by the union of its elements, was a question upon which the greatest chemists entertained different opinions.

The Bakerian Lecture, read before the Royal Society on the 20th of November 1806, not only set this question for ever at rest, but unfolded the mysteries of general Voltaic action; and, as far as theory goes, may almost be said to have perfected our knowledge of the chemical agencies of the pile.

This grand display of scientific light burst upon Europe like a splendid meteor, throwing its radiance into the deepest recesses, and opening to the view of the philosopher new and unexpected regions.

I shall endeavour to offer as popular a review of this celebrated memoir, as the abstruse and complicated nature of its subjects will allow; and I shall be careful in pointing out the successive stages of the enquiry; for we are all too much in the habit of exclusively looking after results; whereas an examination of the steps by which they were attained is far more important, not only to the fame of the discoverer, but to ourselves, as the means of instruction.

The subjects investigated in this memoir are arranged under the following divisions.

1. " On the changes produced in Water by Electricity.

2. " On the agencies of Electricity in the decomposition of various compound Bodies.

3. " On the transfer of certain constituent Parts of Bodies by the action of Electricity.

4. " On the passage of Acids, Alkalies, and other Substances, through various attracting chemical menstrua, by means of Electricity.

5. " Some general Observations on these Phenomena, and on the mode of Decomposition and Transition.

6. " On the General Principles of the chemical changes produced by Electricity.

7. " On the Relations between the Electrical Energies of bodies and their Chemical Affinities.

8. " On the mode of action of the Pile of Volta, with Experimental Elucidations.

9. "On some General Illustrations and Applications of the foregoing facts and principles."

With respect to the first of these divisions, comprehending a history of the changes produced in water by electricity, it is worthy of particular notice, that as early as the year 1800, while residing at Bristol, Davy had discovered that when separate portions of distilled water, filling two glass tubes connected by moist bladders, or any moist animal or vegetable substance, were submitted to the electrical action of the Voltaic pile, by means of gold wires, a *nitro-muriatic* solution of gold appeared in the tube containing the positive wire, and a solution of soda in the opposite tube; but he soon ascertained that the muriatic acid owed its appearance to the animal or vegetable matters employed; for when the same fibres of cotton were used in successive experiments, and washed after every process in a weak solution of nitric acid, the water in the apparatus containing them, though acted upon for a great length of time with a very strong power, produced no effect upon a solution of nitrate of silver.

In every case in which he had procured much soda, the glass* at the point of contact with the wire seemed considerably eroded; when by substituting an agate for a glass cup, no fixed saline matter could be obtained. Its source therefore, in the former case, was evidently the glass.

* It is perhaps a fact not very generally known, that glass, to a certain extent, is decomposable by water: if some of it in a powdered state be triturated with distilled water, in a short time the turmeric test will indicate a portion of alkali in solution.

With respect to Mr. Sylvester's experiment, already noticed, it was sufficient to say that he conducted his process in a vessel of *pipe-clay*, which not only contains lime, but may also include in its composition some of the combinations of a fixed alkali.

On resuming the enquiry, it was Davy's first care to remove every possible source of impurity: he accordingly procured cups of agate, which, previously to being filled, were boiled for several hours in distilled water; and a piece of very white and transparent *amianthus*, a substance first proposed for this purpose by Dr. Wollaston, having been similarly purified, was made to connect the vessels together. Thus was every apparent source of fallacy removed; but still, after having been exposed to Voltaic action for forty-eight hours, the water in the positive cup gave indications of muriatic acid, and that in the negative cup, of soda! The result was as embarrassing as it was unexpected; but it was far from convincing him that the bodies thus obtained were *generated:*—but whence arose the saline matter? Did the agate, after every precaution, still contain some very minute portion of saline matter, not easily discoverable by chemical tests? To determine this question, the experiment was repeated a second, a third, and a fourth time: the quantities of saline matter diminished in every successive operation, which sufficiently proved that the agate must at least have been *one* of the sources sought for; but four additional repetitions of the process convinced the operator that it could not be the only one; that there must exist some other source from which the alkali proceeded, since it continued to appear to the last, in quantities sufficiently distinct, and apparently equal, in every experiment. This was extremely perplexing: every precaution had been taken—the agate cups had even been included in glass vessels, out of the reach of the circulating air—all the acting materials had been repeatedly washed with distilled water; and no part of them in contact with the fluid had ever touched the fingers.

The water itself then, however pure it might appear, must have furnished the alkali. The experiments were repeated in cones of the purest gold, and the water contained in them was submitted to Voltaic action for fourteen hours; the result was, that the acid increased in quantity as the experiment proceeded, and at length became even sour to the taste. On the contrary, the alkaline properties of the fluid in the opposite cone shortly obtained a certain intensity, and remained stationary.

On the application of heat, the alkaline indications became less vivid, although there always remained, after the operation, sufficient evidence to prove that a portion at least was fixed, although probably mixed with ammonia.

The acid, as far as its properties could be examined, agreed with those of pure nitrous acid, having an excess of nitrous gas.

It was now impossible to doubt that the water held in solution some substance which was capable of yielding alkaline matter, but which, from the minuteness of its quantity, had soon been exhausted.

The next step, therefore, was to submit the water

to a still more rigorous examination, which he did by evaporating it in a vessel of silver; when he had the satisfaction to discover the 1-70th of a grain of saline matter.

The water, thus purified in a vessel of silver, was again subjected to Voltaic action in the cones of gold. After two hours, there was only the slightest possible indication of alkali; and this was not, as before, *fixed*, but entirely *volatile*.

In every one of these experiments, acid matter had been produced, and it always presented the character of nitrous acid. Two of the great sources of foreign matter had been detected and removed, viz. the vessels, and the water employed; it still however remained to be explained, how nitrous acid and ammonia could be produced in cases where pure water and pure vessels had been used. In no part of this elaborate enquiry is the penetration of Davy more striking, than in his reasonings upon this problem, and in the beautiful experiments which his sagacity suggested for its solution.

It occurred to him, that the nascent oxygen and hydrogen of the water might respectively combine with a portion of the nitrogen of the common air, which is constantly dissolved in that fluid; but if this were the case, how did it happen that the production of nitrous acid was progressive, while that of the alkali was limited? The experiments of Dr. Priestley, on the absorption of gases by water, at once suggested themselves to his mind as being capable of solving this last difficulty; for that distinguished philosopher had shown, that hydrogen, during its solution in water, expelled the nitrogen, whereas oxygen and nitrogen were capable of co-existing in a state of solution in that fluid. It was, however, necessary to confirm the truth of this explanation by experiment, and he accordingly introduced the two cones of gold, containing purified water, under the receiver of an air-pump; the exhaustion was effected, and the Voltaic pile brought to act upon the water thus circumstanced; after eighteen hours the result was examined, when the water in the negative cone produced no effect upon prepared litmus, but that in the positive vessel did give it a tinge of red barely perceptible.

Had his series of experiments terminated here, the truth of his conclusions would have been established by the comparatively small proportion of acid formed in this latter experiment; but he determined to repeat it under circumstances, if possible, still more unexceptionable and conclusive. Having, therefore, arranged the apparatus as before, he exhausted the receiver, and then filled it with hydrogen gas from a convenient air-holder; he made even a second exhaustion, to ensure the highest accuracy, and then again introduced carefully prepared hydrogen. The Voltaic process was continued during twenty-four hours, and at the end of that period it was found that neither the water in the positive nor in the negative vessels altered the tint of litmus in the slightest degree.

Thus did he succeed in exposing the three great sources of fallacy which had so long misled chemists, with regard to the generation of acid and alkaline matter in Voltaic experiments, viz.—The impurities of the vessels—the foreign matter contained in the water—and the compounds generated by the combination of the nitrogen of atmospheric air with the elements evolved from water; and thus did he establish, by an unbroken chain of incontrovertible evidence, the important truth, that "water, chemically pure, is decomposed by electricity into gaseous matter alone—into oxygen and hydrogen."

Out of the foregoing train of research very naturally sprang the consideration of the *decomposing agencies of Electricity*. It had been constantly observed, that, in all electrical changes connected with the presence of acid and alkaline matter, the former uniformly collected around the positive, and the latter around the negative surface of the apparatus.

In one of the earliest experiments, Davy had also noticed that glass underwent decomposition, and that its alkali always passed to the negative surface. He was, therefore, led to enquire whether, through electrical agency, different solid earthy compounds, insoluble, or soluble with difficulty in water, might not be made to undergo a similar decomposition. We shall find that the results of the trials were decisive and satisfactory. For conducting experiments of this description, he hit upon the happy expedient of constructing the cups with the materials which he wished to submit to experiment, and then by introducing water into them, and forming the necessary connexion by means of asbestus, he completed the Voltaic circuit. In this manner he submitted to experiment *sulphate of lime, sulphate of strontia, fluate of lime, sulphate of baryta*, &c. and with analogous results; the acid element in each case passing to the positive, and the earthy base to the negative cup.

As, in the above experiments, the bodies under examination were presented in considerable masses, and exposed large surfaces to the electric action, it became necessary to enquire whether minute portions of acid and alkaline matter could, by the same agency, be disengaged from solid combinations. This point was very readily elucidated. A piece of fine grained basalt, which, by a previous analysis, had been found to contain 3·5 per cent. of soda, nearly ·5 of muriatic acid, and fifteen parts of lime, having been divided into two properly-shaped pieces, and a cavity, capable of containing twelve grains of water, been drilled in each, was submitted, as in former experiments, to the action of the pile. At the end of ten hours, the result was examined with care, when it appeared that the positively electrified water had the strong smell of oxy-muriatic acid, and copiously precipitated nitrate of silver; while that which was negative affected turmeric, and left by evaporation a residuum which appeared to consist of lime and soda.

A part of a specimen of compact zeolite from the Giants' Causeway, and vitreous lava from Ætna, were each treated in a similar manner, and with results equally satisfactory.

Having thus settled the question with regard to the disengagement of the saline parts of bodies in soluble in water, he proceeded to extend and multiply his experiments on soluble compounds, the decomposition of which, as might have been supposed, always proceeded with greater rapidity, and furnished results more perfectly distinct. In these processes he employed the agate cups, with platina wires, connected by amianthus moistened with pure water; the solutions were introduced into these cups, and the electrifying power applied in the manner already described. In this way, ***sulphate of potash, sulphate of soda, nitrate of potash, phosphate of soda,*** &c. were respectively examined; and in every case the acid, after a certain interval, collected in the cup containing the positive wire, and the alkalies and earths in that containing the negative wire.

When metallic solutions were employed, metallic crystals or depositions were formed on the negative wire, and oxide was likewise deposited around it, while a great excess of acid was found in the opposite cup.

With respect to the transfer of the constituent Parts of Bodies by Electric Action, several original experiments were instituted, and some important conclusions established.

Several facts had been stated, which rendered it probable that the saline elements evolved in decompositions by electricity, were capable of being transferred from one electrified surface to another, according to their usual order of arrangement; but to demonstrate this clearly, farther researches were required, and Davy proceeded to supply the necessary evidence. He connected one of the cups of sulphate of lime before mentioned, with a cup of agate, by means of asbestus, and filling them with purified water, connected them with the battery. In about four hours, a strong solution of lime was found in the agate cup, and sulphuric acid in the cup of sulphate of lime. By reversing the order of arrangement, and carrying on the process during a similar period, the sulphuric acid appeared in the agate cup, and the lime in the opposite vessel. In both these experiments (the acid in the one case, and the lime in the other), the elements of the substance must have passed, in an imperceptible form, along the connecting line of asbestus into the opposite vessel.

Many trials were made with other saline bodies, and with results equally satisfactory; the base always passing into the vessel rendered negative, and the acid into that which was positive.

The time required for these transmissions appeared to be, *cæteris paribus*, in some proportion to the length of the intermediate volume of water.

In the farther prosecution of the enquiry, Davy discovered a still more extraordinary series of facts. In the first place, he found that the contact of the saline solution with a metallic surface was not in the least necessary for its decomposition. He introduced purified water into two glass tubes, and connected with them, by means of amianthus, a vessel containing a solution of muriate of potash. In this case, the saline matter was distant from each of the wires at least two-thirds of an inch; and yet alkaline matter soon appeared in one tube, and acid matter in the other; and in sixteen hours moderately strong solutions of potash and muriatic acid had been formed.

The discovery of this fact became the key to that of others. He very naturally proceeded to enquire into the progress of the transfer, and into the course of the acid and alkaline elements; when, by the use of litmus and turmeric, he arrived at the following conclusion,—that acids and alkalies, during their electrical transference, passed through water containing vegetable colours without effecting in them any change. From which we are led to the consideration of the fourth division of the subject, viz. "On the Passage of Acids, Alkalies, and other Substances, through various attracting Chemical Menstrua, by Electricity."

As soon as it was discovered that a power generated by the Voltaic pile was capable of destroying elective affinity in the vicinity of the metallic points, it seemed reasonable to suppose, that the same power might also destroy it, or at least suspend its operation, throughout the whole of the circuit. The truth of such a supposition was at once placed beyond all doubt by the following very striking experiment.

Three tubes, the first containing a solution of *sulphate of potash*, the second a weak solution of *ammonia*, and the third, *pure water*, each being connected with the other in the usual manner by amianthus, were arranged in relation to the pile, as follow:—the *sulphate of potash* was placed in contact with the negatively electrified point, the *pure water* with the positively electrified point, while the solution of *ammonia* was made the middle link of the conducting chain; so that no sulphuric acid could pass to the positive point in the distilled water, without passing through the ammoniacal solution.

In less than five minutes after the electric current

had been completed, it was found, by means of litmus paper, that acid was in the act of collecting around the positive point; and in half an hour the result was sufficiently distinct for accurate examination.

Other experiments were made with a solution of lime, and with weak solutions of potash and soda, and the results were analogous. Muriatic acid, from muriate of soda, and nitric acid, from nitrate of potash, were also transmitted through concentrated alkaline menstrua, under similar circumstances, and with like effects.

Davy also made several experiments on the transition of alkaline and acid matter, through different neutro-saline solutions, the results of which were exactly such as theory would have anticipated.

In conducting, however, these experiments of electrical transference, there would appear to be one condition essential to their success, viz. that the solution contained in the intermediate vessel should not be capable of forming an insoluble compound with the substance transmitted through it: thus, for example, Davy found that *strontia* and *baryta* passed, like the other alkaline substances, very readily through muriatic and nitric acids; and *vice versâ*, that these acids passed with equal facility through aqueous solutions of the earths in question; but when it was attempted to pass *sulphuric* acid through the same earthy solutions, or to pass the earths through the sulphuric acid, that then the results were of a very different character: the sulphuric acid, in its passage through the barytic solution, was arrested in its progress by the earthy body and falling down as an insoluble compound with it, was carried out of the sphere of the electrical action, by which the power of transfer was destroyed. The same phenomena occurred whenever he attempted to pass muriatic acid through a solution of sulphate of silver. We now come to the next division—viz. "Some general Observations on these Phenomena, and on the mode of Decomposition and Transition."

Davy considers that it will be a general expression of the facts relating to the changes and transitions by electricity, to say, that "hydrogen, the alkaline substances, the metals, and certain oxides, are attracted by negatively electrified, and repelled by positively electrified metallic surfaces; and on the contrary, that oxygen and acid substances are attracted by positively electrified, and repelled by negatively electrified metallic surfaces." And moreover, that these "attractive and repulsive forces are sufficiently energetic to destroy or suspend the usual operation of elective affinity."

Amidst all these wonderful phenomena, that perhaps which excites our greatest astonishment is the fact of the transfer of ponderable matter to a considerable distance, through intervening substances, and in a form that escapes the cognizance of our senses! Upon this question, Davy offers the following remarks:—"It is," says he, "very natural to suppose, that the repellent and attractive energies are communicated from one particle to another particle of the same kind, so as to establish a conducting chain in the fluid; and that the locomotion takes place in consequence: thus, in all the instances in which I examined alkaline solutions through which acids had been transmitted, I always found acid in them, as long as any acid matter remained at the original source. In time, by the attractive power of the positive surface, the decomposition and transfer undoubtedly become complete; but this does not affect the conclusion. In cases of the separation of the constituents of water, and of solutions of neutral salts forming the whole of the chain, there may possibly be a succession of decompositions and recompositions throughout the fluid."

We are next brought to a very important point in the enquiry — viz. "The consideration of the General Principles of the chemical changes produced by Electricity."

The experiment of Mr. Bennett, already alluded to, had shown that many bodies, when brought into contact, and afterwards separated from each other, exhibited signs of opposite states of electricity: but it is to the investigations of M. Volta that we are indebted for the clear developement of the fact; for he has distinctly proved it in the case of copper and zinc, and other metallic combinations, and he supposed that it might also take place with regard to metals and fluids.

In a series of experiments, made in the year 1801, on the construction of electrical combinations, by means of alternations of single metallic plates, and different strata of fluids, as explained upon a former occasion,* Davy had observed that, when acid and alkaline solutions were employed as the elements of these Voltaic combinations, the alkaline solutions

* Page 223.

always received the electricity from, and the acid always transmitted it to the metal. These principles seem to bear an immediate relation to those general phenomena of decomposition and transfer, which have been the subject of the preceding details.

In the most simple case of electrical action, the alkali which receives electricity from the metal would necessarily, on being separated from it, appear *positive*; whilst the acid, under similar circumstances, would be *negative*; and these bodies having respectively, with regard to the metal, that which may be called a positive and a negative electrical energy, in their repellent and attractive functions, would seem to be governed by the common laws of electrical attraction and repulsion; the body possessing the positive energy being repelled by positively electrified surfaces, and that possessing the negative influence following the contrary order.

Davy made a number of experiments with the view of elucidating this idea, and of extending its application; and, in all cases, their results tended, in a most remarkable manner, to confirm the analogy.

He proceeded, by means of very delicate instruments, to ascertain the electrical states of single insulated acid and alkaline solutions, after their contact with metals; but the sources of errors were so numerous, as to render the results far from being satisfactory; but in experiments on dry and solid bodies, the embarrassments arising from evaporation, chemical action, &c. did not occur. When perfectly dry oxalic, succinic, benzoic, or boracic acid, either in the form of powder or crystals, were touched upon an extended surface with a plate of copper, insulated by a glass handle, the copper was found positive, the acid negative. When again metallic plates were made to touch dry lime, strontia, or magnesia, they became negative: in these latter experiments the effect was exceedingly satisfactory and distinct; a single contact upon a large surface being sufficient to communicate a considerable charge.

Numerous other trials were made, and the results confirmed the principle; and moreover proved, as might have been expected, that bodies possessing electrical conditions with regard to one and the same body, possessed them with regard to each other: for instance, a dry piece of lime became positively electrical by repeated contact with crystals of oxalic acid.

These results led him to reason more fully upon the "Relations between the Electrical energies of bodies and their Chemical affinities."

As the chemical attraction subsisting between two bodies seems to be destroyed by giving to one of them an electrical condition opposite to that which it naturally possesses; and since the substances that combine chemically, as far as can be ascertained, exhibit opposite states of electricity, the relations between this energy and chemical affinity would appear to be sufficiently evident to warrant the conclusion at which Davy arrived, viz. that "the combinations and decompositions by electricity were referable to the law of electrical attractions and repulsions;" from which he advanced to the still more important step—"that chemical and electrical attractions were produced by the same cause, acting in one case on particles, in the other on masses."

From these views, he is led to propose the electrical powers, or the forces required to disunite the elements of bodies, as a test or measure of the intensity of chemical attraction. An accurate investigation into this connexion, which may be called the *Electro-dynamic* relations of bodies to their combining masses or proportional numbers, would be the first step towards fixing the science of Chemistry on the permanent foundation of the Mathematics.

If, then, the power of electrical attraction and repulsion be identified with chemical affinity, or rather, if both be dependent upon the same cause, it will follow that two bodies which are naturally in opposite electrical states, may have these states sufficiently exalted to give them an attractive power superior to the cohesive force opposed to their union; when a combination will take place which will be more or less energetic, as the opposed forces are more or less equally balanced. Again, when two bodies, repellent of each other, act upon a third with different degrees of the same electrical energy, the combination will be determined by the degree; or, if bodies having different degrees of the same electrical energy with respect to a third, have likewise different energies with respect to each other, there may be such a balance of attracting and repelling forces as to produce a triple compound; and by the extension of this reasoning, complicated chemical union may be easily explained.

Whenever bodies brought by artificial means into a high state of opposite electricities are made to restore the equilibrium, heat and light are the common consequences. It is perhaps an additional circumstance in favour of the theory to state, that heat and light are likewise the results of all intense chemical action. And as in certain forms of the Voltaic battery, where large quantities of electricity of low intensity act, heat without light is produced; so in slow chemical combinations there is an increase of temperature without any luminous appearance.

The effect of heat in producing combination may be easily explained according to these ideas; it not only gives more freedom of motion to the particles, but in a number of cases it seems to exalt the electrical energies of bodies:—glass, the tourmaline, sulphur, and some others, afford familiar instances of this latter species of energy.

In general, when the different energies are strong and in perfect equilibrium, the combination ought to be quick, the heat and light intense, and the new compound in a neutral state. This would seem to be the case in the combination of oxygen and hydrogen, which form water, a body apparently neutral in electrical energy to most others; and also in the circumstances of the union of the strong alkalies and acids. But where one energy is feeble, and the other strong, all the effects must be less vivid; and the compound, instead of being neutral, ought to exhibit the excess of the stronger energy.

The grand principle thus developed may enable us to obtain new and useful indications of the composition of bodies, by ascertaining the character of their electrical energies; and we now find, in most modern works of Chemistry, that bodies are arranged according to their natural electrical relations; and are said to be ELECTRO-POSITIVE, or ELECTRO-

NEGATIVE, according to their polarities. The advantage of such an arrangement must be freely acknowledged, for it has been the means of establishing analogies * of the utmost importance in chemistry, of which I shall adduce some striking examples in a subsequent part of the present work, when I shall endeavour to offer a general view of the revolution which chemical science has undergone during the investigations of Davy, and contemporary philosophers.

After some further enquiries into the theory of the Voltaic pile,† to which an allusion has been already made, the author offers additional reasons for supposing the decomposition of the chemical menstrua essential to the continued electro-motion of the pile; and if the fluid medium could be a substance incapable of decomposition, there is every reason to believe the equilibrium would be restored, and the motion of the Electricity cease. Having shown the effects of *induction*, in increasing the electricity of the opposite plates, he arrives at the important conclusion, that in a Voltaic arrangement the *intensity of the Electricity increases with the number, but the quantity with the size of the plates.* A theory which was subsequently confirmed by the experiments of Mr. Children.

The paper concludes with "some general illustrations and applications of the foregoing facts and principles," and which the author thinks will readily suggest themselves to the philosophical enquirer.

* It will be sufficient for my present purpose to point out those existing between *Chlorine*, *Iodine*, and *Bromine*.

† See page 226.

They offer, for instance, very easy methods of separating acid and alkaline matter, where they exist in combination in mineral substances; and, in like manner, they suggest the application of electrical powers for effecting the decomposition of animal and vegetable bodies.

On exposing a piece of muscular fibre to the action of the battery, he found that potash, soda, ammonia, lime, and oxide of iron, were evolved on the negative side, and the three mineral, together with the phosphoric, acids, were given out on the positive side.

A laurel leaf, similarly treated, yielded to the negative vessel resin, alkali, and lime; while in the positive one there collected a clear fluid, which had the smell of peach-blossoms, and which, when neutralized by potash, gave a blue-green precipitate to a solution of sulphate of iron; so that it must have contained *Prussic Acid.*

A small plant of mint, in a state of healthy vegetation, on being made the medium of connection in the battery, yielded potash and lime to the water negatively electrified, and acid to that positively electrified. The plant recovered after the process; but a similar one, that had been electrified during a longer period, faded and died.

These facts would seem to show, that the electrical powers of decomposition even act upon vegetable matter in its living condition; and phenomena are not wanting to show that they operate also on the system of living animals. When the fingers, after having been carefully washed with pure water, are brought in contact with this fluid in the positive part of the circuit, acid matter is rapidly developed, having the character of a mixture of muriatic, phosphoric, and sulphuric acids; and if a similar trial be made in the negative part, fixed alkaline matter is as quickly developed.*

Davy thinks that the acid and alkaline taste produced upon the tongue during galvanic experiments, depends upon the decomposition of the saline matter contained in the living animal substance, and perhaps in the saliva; and he farther observes that, as acid and alkaline substances are thus evidently capable of being separated from their combinations in living systems by electrical powers, there is reason to believe that, by converse methods, they might also be introduced into the animal economy, or made to pass through the animal organs; and the same thing may be supposed of metallic oxides; and that these ideas ought to lead to some new investigations in Medicine and Physiology

He thinks it by no means improbable, that the electrical decomposition of the neutral salts, in different cases, may admit of economical applications; and that well-burnt charcoal and plumbago, or charcoal and iron, might be made the exciting powers for such a purpose. Such an arrangement, if erected upon a scale sufficiently extensive, with the medium of a neutro-saline solution, would, in his opinion, produce large quantities of acids and alkalies with very little trouble or expense.

* Reflecting upon this and similar facts, it has occurred to me that Voltaic electricity might be applied for removing the blue colour in the skin, occasioned by the internal use of nitrate of silver. I hope to be able very shortly to submit this theory to the test.

Alterations in chemical equilibrium are constantly taking place in Nature, and he thinks it probable that the electric influence, in its faculties of decomposition and transference, may considerably interfere with the chemical changes occurring in different parts of our system.

The electrical appearances which precede earthquakes and volcanic eruptions, and which have been described by the greater number of the observers of these awful events, admit also of easy explanation on the principles that have been stated.

Besides the cases of sudden and violent change, he considers there must be constant and tranquil alterations, of which electricity, produced in the interior strata of the globe, is the active cause: thus, where *pyritous* strata and strata of *coal-blende* occur, —where the pure metals or the sulphurets are found in contact with each other, or with any conducting substances,—and where different strata contain different saline menstrua, he thinks electricity must be continually manifested; and it is probable that many mineral formations have been materially influenced, or even occasioned, by its agencies.

In an experiment which he performed of electrifying a mixed solution of the muriates of iron, copper, tin, and cobalt, contained in a positive vessel, all the four oxides passed along the connecting asbestus into a positive vessel filled with distilled water, while a yellow metallic crust formed on the wire, and the oxides arranged themselves in a mixed state around the base of it.

In another experiment, in which carbonate of copper was diffused through water in a state of minute division, and a negative wire was placed in a small perforated cube of zeolite in the water, green crystals collected round the cube; the particles not being capable of penetrating it.

By a multiplication of such instances, Davy remarks, that the electrical power of transference may be easily conceived to apply to the explanation of some of the principal and most mysterious facts in geology;* and by imagining a scale of feeble powers, it would be easy to account for the association of the insoluble metallic and earthy compounds containing acids.

"Natural electricity," observes our philosopher, "has hitherto been little investigated, except in the case of its evident and powerful concentration in the atmosphere. Its slow and silent operations in every part of the surface will probably be found more immediately and importantly connected with the order and economy of nature; and investigations on this subject can hardly fail to enlighten our philosophical systems of the earth, and may possibly place new powers within our reach."

Thus concludes one of the most masterly and powerful productions of scientific genius. I may perhaps have been considered prolix in recording the progressive researches by which he arrived at his results; but let it be remembered, that the great fame of Davy, as an experimental philosopher, rests

* During the contentions of the Neptunists and Plutonists, alluded to in a former part of this work, specimens were produced exhibiting the intermixture of mineral bodies, which was completely hostile to all theory. These anomalies now receive a plausible explanation from the agencies of Voltaic electricity.

upon this single memoir; and though the secondary results to be hereafter considered, may be more dazzling to ordinary minds, yet in the judgment of every scientific observer, they must appear far less glorious than the discovery of the primitive laws. Let me ask whether Sir Isaac Newton does not deserve greater fame for his invention of fluxions, than for the calculations performed by the application of them? I do not hesitate in comparing these great philosophers, since each has enlightened us by discoveries alike effected by means invented by himself. Not only did both unlock the caskets of Nature, but they had the superior merit of planning and constructing the key.

I challenge those, who have carefully followed me through the details of the preceding memoir, to show a single instance in which accident, so mainly contributory to former discoveries in electricity, had any share in conducting its author to truth. Step by step did he, with philosophic caution and unwearied perseverance, unfold all the particular phenomena and details of his subject; his genius then took flight, and with an eagle's eye caught the plan of the whole.—A new science has been thus created; and so important and extensive are its applications, so boundless and sublime its views, that we may fairly anticipate the fulfilment of those prophetic words of Dr. Priestley, who, in the preface to his History of Electricity,* exclaims—"Electricity seems to be giving us an inlet into the internal

* The History and Present State of Electricity; by Jos. Priestley, LL.D. F.R.S., &c. London, 1795.

structure of bodies, on which all their sensible properties depend. By pursuing this new light, therefore, the bounds of natural science may possibly be extended beyond what we can now form any idea of. New worlds may open to our view, and the glory of the great Sir Isaac Newton himself, and all his contemporaries, be eclipsed by a new set of philosophers, in quite a new field of speculation. Could that great man revisit the earth, and view the experiments of the present race of electricians, he would be no less amazed than Roger Bacon, or Sir Francis, would have been at his." In our turn we may ask, what would be the astonishment—what the delight of Dr. Priestley, could he now witness the successful results of Voltaic research?—and what would he say of that mighty genius who has demonstrated the relations of electrical energy to the general laws of chemical action?* It was his good fortune to have witnessed the discovery which identified electricity with the lightning of the thunder cloud: what would he have said of that which identified it with the magnetism of the earth! Of this at least we may be certain, that he would have expunged from his history the passage in which he observes—"Electrical discoveries have been made so much by accident, that it is more the powers of nature, than of human genius, that excite our wonders with respect to them."

* Dr. Priestley augured much from the talents of Davy. After the publication of his first paper on Galvanism, he wrote to him from America, and expressed the pleasure he felt on finding his favourite subject in such able hands. Priestley died in 1804, and therefore did not witness Davy's success.

CHAPTER VII.

The unfair rivalry of Philosophers.—Bonaparte the Patron of Science. — He liberates Dolomieu. — He founds a Prize for the encouragement of Electric researches.—His letter to the Minister of the Interior. — Proceedings of the Institute. — The Prize is conferred on Davy. — The Bakerian Lecture of 1807. — The Decomposition of the Fixed Alkalies—Potassium—Sodium.—The Questions to which the discovery gave rise.—Interesting Extracts from the Manuscript Notes of the Laboratory.—Potash decomposed by a chemical process.—Letters to Children, and Pepys.—The true nature of Potash discovered.—Whether Ammonia contains Oxygen.—Davy's severe Illness.—He recovers and resumes his labours.—His Fishing Costume.—He decomposes the Earths. — Important views to which the discovery has led.

It must be confessed that there has too frequently existed amongst philosophers a strange and ungenerous disposition to undervalue the labours of their contemporaries. If a discovery be made, its truth and importance are first questioned; and should these be established, then its originality becomes a subject of dispute.

Truth, although she may have been rarely held fast, has been frequently touched* in the dark: it

* A most remarkable illustration of this fact occurs in the history of Locke, who certainly came as near to an important dis-

is not extraordinary, therefore, that evidence may be often strained from the writings of philosophers in support of prior claims to late discoveries; but upon a candid review, these loose statements, or obscure hints, will generally be found wholly destitute of the pretensions which an unfair spirit of rivalry has too often laboured to support. Many of such hints, indeed, so far from advancing the progress of truth, had never even attracted notice, until after the discoveries to which they have been supposed to relate.

covery as any philosopher who ever caught a glimpse of a truth without seizing it; but his statement did not, in any degree, hasten the developement of that new branch of science which was reserved for the genius of Dr. Black to investigate, and who a century later, by the discovery of fixed air, changed the whole face of Chemistry. The passage to which I allude is extracted from the Life by Lord King, and is so curious, that I shall give it a place in this note. "M. Toinard produced a large bottle of Muscat: it was clear when he set it on the table; but when he had drawn out the stopper, a multitude of little bubbles arose, and swelled the wine above the mouth of the bottle. It comes from this, that the air, which was included and disseminated in the liquor, had liberty to expand itself, and so to become visible, and, being much lighter than the liquor, to mount with great quickness.—*Quere,* Whether this be air new generated, or whether the springy particles of air in the fruit, out of which these fermenting liquors are drawn, have, by the artifice of Nature, been pressed close together, and there by other particles fastened and held so; and whether fermentation does not loose these bands, and give them liberty to expand themselves again? Take a bottle of fermenting liquor, and tie a bladder on the mouth.—*Quere,* How much new air will it produce; and whether this has the quality of common air?"

Another instance equally illustrative of the manner in which important truths will sometimes elude notice, even after Science has approached so near as to touch them, is presented in the history of the Barometer. Toricelli, the pupil of Galileo, while

Although the importance of Davy's Electro-chemical discoveries could not for a moment be doubted; their claims to originality, it would seem, were not admitted without some question. The works of Ritter and Winterl, amongst many others, were quoted to show that these philosophers had imagined or anticipated the relation between electrical powers and chemical affinities; but Davy very fairly observes, in a paper read before the Royal Society in 1826, that in the obscurity of the language and metaphysics of both those gentlemen, it is difficult

reflecting upon the phenomenon which had so greatly perplexed his master, viz. that water could not be raised above thirty-two feet in the body of a pump, rightly conjectured that the water, under such circumstances, was not *drawn*, but *pushed up* into the barrel, and that it could only be so pushed up by the force ofthe atmosphere. It then occurred to him, that if mercury were used instead of water, being heavier, it would not be pushed up so high by the weight of the air So, taking a glass tube of about three feet in height, made air-tight at one end, he first filled it completely with quicksilver, and then closing it with his finger, reversed it in a basin containing that metal; when he had the gratification of seeing the liquid in the tube descend, as he had anticipated. Here then was the discovery of the BAROMETER; but it was reserved for another to find out that such an instrument had been actually invented. Pascal first made the remark, that the inference of Toricelli, if true, might be confirmed by carrying the mercurial tube to a considerable elevation; when the atmospheric column being diminished, that of the mercury, which was supposed to be its balance, ought likewise to be shortened in a corresponding proportion. It followed then, that a measure of the weight of the atmosphere, in all circumstances, had been obtained, and consequently that of the height of any place to which the instrument could be carried. In this manner was a discovery completed, which had for ages escaped the greatest philosophers who had made the nearest approaches to its developement.

to say what may not be found. In the ingenious though wild views of Ritter, there are hints which may more readily be considered as applying to *Electro-magnetism* than to *Electro-chemistry*; while Winterl's *Miraculous Andronia* might, with as much propriety, be considered as a type of all the chemical substances that have been since discovered, as his view of the antagonist powers (the acid and base) be regarded as an anticipation of the *Electro-chemical* theory.

It would be worse than useless to speak of other works, which refer the origin of Electro-chemistry to Germany, Sweden, and France, rather than to Italy and England; and which attribute some of the views first developed by Davy, to philosophers who have not, nor ever could have made any claim of the kind, since their experiments were actually not published until many years after 1806, the date of the Bakerian Lecture.

With regard to the judgment of posterity upon these points, but little apprehension can be entertained. I well remember, in a conversation with Davy, he observed, that "a philosopher might generally discover how his labours would be appreciated in after ages, from the opinion entertained of them by contemporary foreigners, who, being unbiassed by circumstances of personality, will reduce every object to its just proportions and value."

If we acknowledge the truth of such a standard, and submit the posthumous fame of Davy to its measure, where is the philosopher, in our times, whose name is destined to attain a higher eminence in the history of Science? Let the reader only recall to his recollection the bitter animosity which France and England mutually entertained towards each other in the year 1807, and he will be able to form some idea of the astounding impression which the Bakerian Lecture must have produced on the Savans of Paris, when, in despite of national prejudice and national vanity, it was crowned by the Institute of France with the prize of the First Consul! Thus did the Voltaic battery, in the hands of the English chemist, achieve what all the artillery of Britain could never have produced—A SPONTANEOUS AND WILLING HOMAGE TO BRITISH SUPERIORITY!—But let not this observation convey the slightest idea of disrespect, or be supposed to encourage any feeling to the disparagement of the chemists of France; on the contrary, it is even a question not readily answered, to which party the triumph fairly belongs,—to him who won the laurel crown, or to those who so nobly placed it on his brow? They have set an example to future ages, which may as materially advance the progress of science, as the researches which called it forth:—they have shown, to adopt the language of an eloquent writer, that "the Commonwealth of Science is of no party, and of no nation; that it is a pure Republic, and always at peace. Its shades are disturbed neither by domestic malice nor foreign levy; they resound not with the cries of faction or of public animosity. Falsehood is the only enemy their inhabitants denounce; Truth, and her minister Reason, the only leaders they follow."

I shall avail myself of this opportunity to introduce the Report drawn up by M. Biot, and made in the name of a Commission appointed by the Institute to accomplish the intention of Bonaparte, who, when First Consul, founded prizes for important discoveries in Electricity or Galvanism.

It is an opinion very generally received, that despotism is hostile to the progress of Philosophy—that the suspicion natural to tyranny, and the fear that light should expose its deformity, have, under such circumstances, inspired a dread of any thing approaching to freedom of enquiry. The conduct of Napoleon, not only during his Consulate, but even after he had assumed the Purple, is in direct opposition to such an opinion. Now that the excitements of national hostility have subsided, and the asperity of our feelings towards that extraordinary man has been softened by time and prosperity, we are enabled to discern the bright and sunny spots in his character.

Not to mention the immense plans which his genius suggested for the internal improvement of France, the annals of the Institute would furnish innumerable proofs of the zeal with which he encouraged Science, and promoted its interests.

His liberation of Dolomieu from the dungeons of Tarentum was an act not only remarkable for the considerate regard it displayed for Science, but for the spirit and eagerness with which it was effected. The French Government had repeatedly made the most urgent demands for the liberty of one who had reflected so much credit on his country;—the Danes had also directed the interference of their Minister, and the King of Spain had added his solicitations in vain:—no sooner, however, had the astonishing campaign which terminated by the victory of Marengo, completely established the French Republic, than Bonaparte, in making peace with Naples, stipulated for the immediate deliverance of Dolomieu, as the first article of the treaty.

The following letter from Bonaparte, addressed to the Minister of the Interior, and by him transmitted to the Institute, expresses the intentions of the First Consul, in founding prizes for important discoveries in Electricity or Galvanism.

"I intend, Citizen Minister, to found a prize, consisting of a Medal of three thousand francs, (about one hundred and twenty pounds sterling,) for the best experiment which shall be made in the course of each year, on the Galvanic fluid.

"For this purpose, the Memoirs containing the details of the said experiments shall be sent before the First of *Fructidor*, to the Class of the Mathematical and Physical Sciences, which in the complimentary days shall adjudge the prize to the author of that experiment which has been most useful to the progress of Science.

"I also desire to give, by the way of encouragement, the sum of sixty thousand francs to the person who, by his experiments and discoveries, shall, according to the opinion of the Class, advance the knowledge of Electricity and Galvanism as much as Franklin and Volta did.*

* "À celui qui, par ses expériences et ses découvertes, *fera à faire à l'Electricité et au Galvanisme un pas comparable à celui qu'ont fait faire à ces Sciences Franklin et Volta.*"

My French correspondent adds, "Ces soixantes mille francs n'ont pas été adjugés, *le pas n'ayant point été fait.*"

"Foreigners of all nations are admitted to the competition.

"I beg you will make known these dispositions to the President of the First Class of the National Institute, that it may give to these ideas such developement as may appear proper; my particular object being to encourage philosophers, and to direct their attention to this part of philosophy, which, in my opinion, may lead to great discoveries.

"(Signed) BONAPARTE."

Upon the presentation of this letter, a Committee was appointed to consider the means for accomplishing the intentions of the First Consul; and after expatiating upon the extensive agencies of Electricity, their Report concludes in the following manner:—

"To fulfill the intention of the First Consul, and to give to the competition all the solemnity which the importance of the object, the nature of the Prize, and the character of the Founder require, the Commissioners unanimously propose as follows:

"The Class of the Mathematical and Physical Sciences of the National Institute opens the general competition required by the First Consul.

"All the learned of Europe, and the Members and Associates of the Institute, are admitted to the competition.

"The Class does not require that the Memoirs should be immediately addressed to it. Every year it will crown the author of the best experiments which shall come to its knowledge, and which shall have advanced the progress of the science.

"The present report, containing the letter of the

First Consul, shall be printed, and serve as a programme.

"Done at the National Institute, Messidor 11, year 10.

"(Signed) LAPLACE, HALLE, COULOMB, HAUY. BIOT, Reporter."

It was not until twelve months after the publication of his first Bakerian Lecture, that Davy received the intelligence that the prize of three thousand francs had been awarded him by the Institute of France, for his discoveries announced in the Philosophical Transactions for the year 1807.

Mr. Poole, in a late communication, informs me that he was in London soon after the letter communicating this gratifying intelligence had been received from France; and that Davy, upon showing it to him, observed, "Some people say I ought not to accept this prize; and there have been foolish paragraphs in the papers to that effect; but if the two countries or governments are at war, the men of science are not. That would, indeed, be a civil war of the worst description: we should rather, through the instrumentality of men of science, soften the asperities of national hostility."

After Davy had been elected Secretary to the Royal Society, he appears to have been confined to town during the autumn of 1807, when he wrote the following letter.

TO THOMAS POOLE, ESQ.

MY DEAR POOLE, August 28th, 1807.

I AM obliged to be in the neighbourhood of town during the greater part of the summer, for the purpose of correcting the press for the Philosophical Transactions.

I made a rapid journey into Cornwall for the sake of seeing my family; and it was not in my power, had I received your letter at Lyme, to have accepted your kind invitation.

If C—— is still with you, will you be kind enough to say to him, that I wrote nearly a week ago two letters about lectures, and not knowing where he was, I addressed them to him at different places? I wish very much he would seriously determine on this point. The Managers of the Royal Institution are very anxious to engage him; and I think he might be of material service to the public, and of benefit to his own mind, to say nothing of the benefit his purse might also receive. In the present condition of society, his opinions in matters of taste, literature, and metaphysics, must have a healthy influence; and unless he soon become an actual member of the living world, he must expect to be hereafter brought to judgment 'for hiding his light.'

The times seem to me to be less dangerous, as to the immediate state of this country, than they were four years ago. The extension of the French Empire has weakened the disposable force of France. Bonaparte seems to have abandoned the idea of invasion; and if our Government is active, we have little to dread from a maritime war, at least for some time. Sooner or later, our Colonial Empire must fall in due time, when it has answered its ends.

The wealth of our island must be diminished, but the strength of mind of the people cannot easily pass away; and our literature, our science, our arts, and the dignity of our nature, depend little upon our external relations. When we had fewer colonies than Genoa, we had Bacons and Shakspeares.

The wealth and prosperity of the country are only the *comeliness* of the body—the fulness of the flesh and fat;—but the spirit is independent of them; it requires only muscle, bone, and nerve, for the true exercise of its functions. We cannot lose our liberty, because we cannot cease to *think;* and ten millions of people are not easily annihilated.

I am, my dear Poole, very truly yours,

H. DAVY.

While the Electro-chemical laws, developed in the last chapter, are fresh in the recollection of the reader, I shall proceed to the consideration of his second Bakerian Lecture, which was read in November 1807; and in which he announces the discovery of the metallic bases of the fixed alkalies,—a discovery immediately arising from the application of Voltaic electricity, directed in accordance with those laws;—thus having, as we have seen in the first instance, ascended from particular phenomena to general principles, he now descends from those principles to the discovery of new phenomena: a method of investigation by which he may be said to have applied to his inductions the severest tests of truth, and to have produced a chain of evidence without having a single link deficient.

Since the account given by Newton of his first discoveries in Optics, it may be questioned whether so happy and successful an instance of philosophical induction has ever been afforded as that by which Davy discovered the composition of the fixed alkalies. Had it been true, as was most unjustly insinuated at the time, that the discovery was accidentally effected by the high power of the apparatus placed at his disposal, his claims to our admiration would have assumed a very different character: in such a case, he might be said to have forced open the sanctuary of Nature by direct violence, instead of having discovered and touched the secret spring by which its portals were unclosed. The justice of these remarks will best appear in the examination of his memoir: the highest eulogy that can be conferred on its author will be a faithful and plain history of its contents.

It will be remembered that, in his preceding lecture of 1806, he had described a number of decompositions and chemical changes produced in substances of known composition, by the powers of electricity, and that in all such cases there invariably subsisted an attraction between oxygen and the *positive* pole, and between inflammable matter and the *negative* pole of the pile: thus, in the decomposition of water, its oxygen was transferred to the former, and its hydrogen to the latter. Furnished with such data, Davy proceeded to submit a fixed alkali to the most intense action of the Voltaic apparatus, well convinced that, should the electrical energy be adequate to effect its decomposition, the elements would be transferred, according to this general law, to their respective poles.

His first attempts were made on solutions of the alkalies; but, notwithstanding the intensity of the electric action, the water alone underwent decomposition, and oxygen and hydrogen were disengaged with the production of much heat, and violent effervescence. The presence of water thus appearing to prevent the desired decomposition, potash, in a state of igneous fusion, was in various ways submitted to experiment; when it was evident that combustible matter of some kind, burning with a vivid light, was given off at the negative wire. After numerous trials, during the progress of which the difficulties which successively arose were as immediately combated by ingenious manipulation, a small piece of potash sufficiently moistened, by a short exposure to the air, to give its surface a conducting power, was placed on an insulated disc of platina, connected with the negative side of the battery in a state of intense activity, and a platina wire communicating with the positive side, was at the same instant brought into contact within the upper surface of the alkali.—Mark what followed!—A series of phenomena, each of which the reader will readily understand as it is announced,—for it will be in strict accordance with the laws which Davy had previously established:—the potash began to fuse at both its points of electrization: a violent effervescence commenced at the upper, or positive surface; while at the lower, or negative one, instead of any liberation of elastic matter, which would probably have happened had hydrogen been an element of the alkaline body, small globules, resembling quicksilver, appeared, some of which were no sooner formed than they burnt with explosion and bright flame.—What must have been the sensations of

Davy at this moment!—He had decomposed potash, and obtained its base in a metallic form.

The gaseous matter developed, during the experiment, at the positive pole of the apparatus, he very shortly identified as oxygen. To collect, however, the metallic matter, in a quantity sufficient for a satisfactory examination, was by no means so easy; for, like the *Alkahest* imagined by the Alchemist, it acted more or less upon every body to which it was exposed; and such was its attraction for oxygen, that it speedily reverted to the state of alkali by recombining with it.

After various trials, however, it was found that recently distilled naphtha presented a medium in which it might be preserved and examined, since a thin transparent film of this fluid, while it defended the metal from the action of the atmosphere, did not oppose any obstacle to the investigation of its physical properties.

Thus provided, he proceeded to enquire into the nature of the new and singular body, to which he afterwards gave the name of POTASSIUM, and which may be described as follows.

Its external character is that of a white metal, instantly tarnishing by exposure to air; at the temperature of 70° Fah. it exists in small globules, which possess the metallic lustre, opacity, and general appearance of quicksilver; so that when a globule of the latter is placed near one of the former, the eye cannot discover any difference between them: at this temperature, however, the metal is not perfectly fluid; but when gradually heated, it becomes more so,—and at 150°, its fluidity is so perfect that several globules may be easily made to run into one. By reducing its temperature, it becomes, at 50°, a soft and malleable solid, which has the lustre of polished silver, and is soft enough to be moulded like wax. At about the freezing point of water it becomes hard and brittle, and exhibits, when broken, a crystallized structure of perfect whiteness, and of high metallic splendour. It is also a perfect conductor both of electricity and heat. Thus far, then, it fulfills every condition of a metal; but an anomaly of a most startling description has now to be mentioned—the absence of a quality which has been as invariably associated with the idea of a metal, as that of lustre, viz. great specific gravity. Whence a question has arisen, whether, after all, the alkaline base can with propriety be classed under that denomination? Instead of possessing that ponderosity which we should have expected in a body otherwise metallic, it is so light as not only to swim upon the surface of water, but even upon that of naphtha, by far the lightest liquid in nature. Davy, however, very justly argues, that low specific gravity does not offer a sufficient reason for degrading this body from the rank of a metal; for amongst those which constitute the class, there are remarkable differences with respect to this quality; that platina is nearly four times as heavy as tellurium. In the philosophical division of bodies into classes, the analogy between the greater number of properties must always be the foundation of arrangement.*

* The propriety, and even the necessity, of such a compact become daily more apparent, as our knowledge of bodies extends. If

So inseparable however, by long association, are the ideas of ponderosity and metallic splendour, that the evidence even of the senses may fail in disuniting them.* This is well illustrated by the following amusing anecdote. Shortly after the discovery of potassium, Dr. George Pearson happened to enter the laboratory in the Royal Institution, and upon being shown the new substance, and interrogated as to its nature, he, without the least hesitation, on seeing its lustre, exclaimed, "Why, it is metallic, to be sure," and then, balancing it on his finger, he added, in the same tone of confidence, "*Bless me, how heavy it is!*"

When thrown upon water, potassium instantly decomposes that fluid, and an explosion is produced

we were to degrade a substance from its class, in consequence of the absence of some one quality which enters into its more perfect examples, we should soon find ourselves involved in paradoxes.—What idea, for instance, could we form of an acid?—Its sourness? —Prussic Acid—Arsenious Acid, are not sour.—Its tendency to combine with an alkaline or earthy base?—If so, sugar is an acid, for it combines with lime. I remember a chemist having been exposed to much ridicule from speaking of a *sweet* acid—Why not?

* In the language of Darwin, we should say, that the simple ideas of weight and lustre, which form the complex idea of a metal, have become so indissoluble, that they can no longer be separated by volition. The principle admits of many familiar illustrations, and is the source of numerous fallacies. When any one voluntarily recollects a Gothic window, which he had seen some time before, the whole front of the Cathedral occurs to him at the same time: in like manner, the taste of a pine-apple, though we eat it blindfold, recalls the colour and shape of it. Coleridge has made a good remark upon this subject. He says, "It is a great law of the imagination, that a likeness in part tends to become a likeness of the whole." It is thus that we trace images in the fire, castles in the clouds, and spectres in the gloom of twilight.

with a vehement flame: an experiment which is rendered more striking if, for water, ice be substituted; in this latter case, it instantly burns with a bright rose-coloured flame, and a deep hole is made in the ice, which will afterwards be found to contain a solution of potash.

It is scarcely necessary to state, that these phenomena depend upon the very powerful affinity which the metal possesses for oxygen, enabling it even to separate it from its most subtle combinations.*

One of the neatest modes of showing the production of alkali, in the decomposition of water by the basis of potash, consists in dropping a globule of potassium upon moistened paper tinged with turmeric. At the moment that it comes into contact with the water, it burns and moves rapidly upon the paper, as if in search of moisture, leaving behind it a deep reddish-brown trace of its progress, and acting upon the test paper precisely as dry caustic potash.

* If we are disposed to enter into a more critical examination of the subject, we shall find that, although the above is a general expression of the change produced, there are subordinate actions of a more complicated nature: the metal, in the first place, decomposes a portion of the water, in order to combine with its oxygen, and form potash, which in its turn has a powerful affinity for water; the heat arising from two causes, decomposition and combination, is sufficiently intense to produce the inflammation. Water is a bad conductor of heat; the globule swims exposed to air; a part of which is dissolved by the heated nascent hydrogen; and this gas, being capable of spontaneous inflammation, explodes and communicates the effect of combustion to any of the bases that may be yet uncombined. The manner in which the potassium runs along the surface of water may be compared to a drop of water on red-hot iron; in the one case the hot potassium, in the other the cold water, is enveloped in an atmosphere of steam.

From these observations, the reader will immediately perceive, that the decomposition of the fixed alkalies has placed in the hands of the experimentalist a new instrument of research, scarcely less energetic, or of less universal application, than the power from which the discovery emanated. Davy observes upon this point, that "it will undoubtedly prove a powerful agent for analysis, and having an affinity for oxygen, stronger than any other known substance, it may possibly supersede the application of electricity to some of the undecompounded bodies." So strong indeed is its affinity for oxygen, that it discovers and decomposes the small quantities of water contained in alcohol and ether; and in the latter case, this decomposition is connected with an instructive result. Potash is insoluble in that fluid: when therefore its base is thrown into it, oxygen is furnished, hydrogen gas disengaged, and the alkali, as it is regenerated, renders the ether white and turbid.

But perhaps the most beautiful illustration of its deoxidizing power is afforded by its action on carbonic acid gas, or fixed air: when heated in contact with that gas, it catches fire, and by uniting with its oxygen, becomes potash, while the liberated carbon is deposited in the form of charcoal.

As I have already exceeded the limits originally prescribed to myself, I shall not enter into the history of Davy's experiments on the other fixed alkali, soda, farther than to state that, when it was submitted to Voltaic action, a bright metal was obtained, similar in its general characters to potassium, but possessing sufficiently distinctive peculiarities as to volatility, fusibility, oxidability, &c. To this body Davy assigned the name of SODIUM.*

In support of the metallic characters of these alkaline bases, it may be necessary to state that they combine with each other, and form alloys; the properties and habitudes of which are very interesting, and are fully described by their discoverer.

No sooner had these results been made known to the scientific world, than a question arose, both in this country and abroad, as to the real nature of the bodies which had been thus obtained from the fixed alkalies, and which presented an aspect so obviously metallic. At first, it was conjectured by a few, that they might be compounds of the alkali with the platina used in the experiments; but this was at once disproved by Davy having obtained the same results when pieces of copper, silver, gold, plumbago, or even charcoal, had been employed for completing the Voltaic circuit.

The effect which this and his subsequent discoveries produced, in revolutionizing the theory of Chemistry, will form an interesting subject for discussion in a future part of the present work: I shall therefore only remark in passing, that the fact of oxygen, the acknowledged principle of acidity, existing in combination with a metallic base, and imparting to it the properties of an alkali, was no sooner announced, than its truth was strenuously denied. It was an attack upon opinions sanctioned

* In his Bakerian Lecture of 1810, he informs us that he obtains Sodium by heating common salt, which has been previously ignited, with Potassium—an immediate decomposition takes place, and two parts of Potassium produce rather more than one of Sodium.

by the general suffrage of the scientific world;—it was, in fact, storming the very citadel of their philosophy: no wonder, then, that the agitator should have been assailed with a full cry for his revolutionary plans.* M. Curadau read a paper before the French Institute, in which he endeavoured to prove, *first*, that the conversion of the alkalies into metals was not a deoxidation of those bodies, but a combination of them with new elements;—*secondly*, that the affinity of the alkaline metals for oxygen was merely a chemical illusion, occasioned by some body the presence of which had not been suspected;—*thirdly*, that carbon was one of the elements of the alkaline metals, since it could be separated from them at pleasure, or converted into carbonic acid;—and *fourthly*, that if the specific gravities of the new substances were less than that of water, it was because hydrogen was associated with carbon in the combination.

It is scarcely necessary to state, that the presence of carbon was readily traced to sources of impurity. The hypothesis which assumed the existence of hydrogen as an element, was not so easily refuted. It was espoused by MM. Gay Lussac, Thenard, and Ritter, on the Continent, and by Mr. Dalton in England. The former derived their inference from the action of potassium upon ammonia, by which

* Many years afterwards, when Davy was travelling on the Continent, a distinguished person about a foreign court, enquired who and what he was; never having heard of his scientific fame. Upon being told that his discoveries had revolutionized Chemistry, the courtier promptly replied—"I hate all revolutionists—his presence will not be acceptable here."

they obtained a fusible substance that yielded by heat more hydrogen than the ammonia contained; the latter contended that potassium and sodium are proved to be *hydrurets*, by the very process employed for their production; for, since common potash is a *hydrat*, and oxygen is produced at one surface, and potassium at the other, by Voltaic action, he conceived that the former arose from the decomposition of water, and that the hydrogen must therefore unite with the potash to form potassium. It is a curious fact, that Berthollet, in the very sentence in which he insisted upon the excessive quantity of hydrogen disengaged in his experiment, as a proof that potassium must be a *hydruret*, should have stated that the addition of water to the residuum was necessary for obtaining his result. How could it have happened that he overlooked so obvious a source of hydrogen? Mr. Dalton, as well as Ritter, considered the low specific gravity as favouring the idea of their containing hydrogen; but Davy observes that they are less volatile than antimony, arsenic, and tellurium, and much less so than mercury. Besides, sodium absorbs much more oxygen than potassium, and, on the hypothesis of hydrogenation, must therefore contain more hydrogen; and yet though soda is said to be lighter than potash, in the proportion of thirteen to seventeen nearly, sodium is heavier than potassium, in the proportion of nine to seven at least. On the theory of Davy, this circumstance is what ought to have been expected. Potassium has a much stronger affinity for oxygen than sodium, and must condense it much more; and the resulting higher specific gravity of the combination is a necessary consequence. In this manner did Davy entangle his opponents in their own arguments, and establish, in the most triumphant manner, the truths of his original views.

Thus then was a discovery effected, and at once rendered complete, which all the chemists in Europe had vainly attempted to accomplish. The alkalies had been tortured by every variety of experiment which ingenuity could suggest, or perseverance perform, but all in vain; nor was the pursuit abandoned until indefatigable effort had wrecked the patience and exhausted every resource of the experimentalist. Such was the disheartening, and almost forlorn condition of the philosopher when Davy entered the field:—he created new instruments, new powers, and fresh resources; and Nature, thus interrogated on a different plan, at once revealed her long cherished secret.

In his Bakerian Lecture, Davy observes, that "a historical detail of the progress of the investigation of all the difficulties that occurred, and of the manner in which they were overcome, and of all the manipulations employed, would far exceed the limits assigned to a Lecture." But to the chemist, every circumstance, however minute, connected with a subject of such commanding importance, is pregnant with interest; I therefore considered it my duty to search into the archives of the Institution, in the hope that I should discover some memoranda which might supply additional information. In examining the Laboratory Register, I have so far succeeded as to obtain some rough and imperfect notes, which will, to a certain degree, assist us in analysing the intellectual operations by which his mind ultimately arrived at the grand conclusion.

It appears from this register that Davy commenced his enquiries into the composition of potash on the 16th, and obtained his great result on the 19th of October 1807 * His first experiments, however, evidently did not suggest the truth: he does not appear to have suspected the nature of the alkaline base until his last experiment, when the truth flashed upon him in the full blaze of discovery. His first note, dated the 16th, leads us to infer that he acted on a solid piece of potash, under the surface of alcohol, and several other liquids in which the alkali was not soluble; and that he obtained gaseous matter, which he called at the moment '*Alkaligen Gas*,' and which he appears to have examined most closely, without arriving at any conclusion as to its nature. On the following day, he, for the first time, would seem to have developed potassium by electric action on potash under oil of turpentine, for the note records the fact of "*the* globules giving out gas by water, which gas *burnt in contact with air*;" and then follows a query—"Does *it*" (the matter of the globules) "not form gaseous compounds with ether, alcohol, and the oils?" Here, then, he evidently imagined, that the matter of the globules, which he had never obtained from potash, except when acted upon under oil of turpentine, had formed gaseous compounds with the ether, alcohol, and oils in his previous experi-

* On the same day he decomposed Soda with somewhat different phenomena.

ments, and given origin to that which he had termed '*Alkaligen Gas*.'

He then leaves the consideration of this gas, and attacks the unknown globules, which probably did not present any metallic appearance under the circumstances in which he saw them, for they must have been as minute as grains of sand. I rather think that he commenced his examination by introducing a globule of mercury, and uniting it with a globule of the unknown substance, for his note says, "Action of the substance on Mercury,—forms with it a solid amalgam, which soon loses its *Alkaligen* in the air." And from the note which succeeds, he evidently considered this *Alkaligen* (potassium) volatile, as he says "it soon flies off on exposure to the air."

October 19.—It is probable that, in consequence of the property which the unknown substance displayed of amalgamating with mercury, he devised his experiment of the 19th. He took a small glass tube, about the size and shape of a thimble, into which he fused a platinum wire, and passed it through the closed end. He then put a piece of pure potash into this tube, and fused it into a mass about the wire, so as entirely to defend it from the mercury afterwards to be used. When cold, the potash was solid, but containing moisture enough to give it a conducting power; he then filled the rest of the tube with mercury, and inverted it over the trough: the apparatus being thus arranged, he made the wire and the mercury alternately positive and negative. And now, conceiving that I have sufficiently explained his brief notes, the reader shall receive the result in his own words: for this purpose I have obtained an engraving of the autograph, which is here annexed; but as it may not be very readily deciphered, I shall first give the substance of it in print.—"When potash was introduced into a tube having a platina wire attached to it—so—and fused into the tube so as to be a conductor, *i. e.* so as to contain just water enough, though solid, and inserted over mercury, when the platina was made negative, no gas was formed, and the mercury became oxydated, and a small quantity of the *alkaligen* was produced round the platina wire, as was evident from its quick inflammation by the action of water. When the mercury was made the negative, gas was developed in great quantities from the positive wire, and none from the negative mercury, and this gas proved to be pure oxygene—A CAPITAL EXPERIMENT, PROVING THE DECOMPOSITION OF POTASH." The Reviewer of the Institution Journal well observes that those who knew Davy will best conceive the enthusiasm with which this hasty record of his success was dashed off, and will instantly recognise ευρηκα in his "CAPITAL EXPERIMENT."

From this same Register, it appears that, in the preceding month, he was deeply engaged in experiments on '*Antwerp blue*,' which he found to consist of ***Prussiate of Iron*** and ***Alumina***, "probably in the proportion of two-thirds of the former to one-third of the latter."

On the 6th of October, we learn from the same source, that he performed a beautiful experiment, that of producing the vegetation of the carbon of the wick of a candle, by placing it between the wires of the battery.

On the 12th of the preceding September he addressed a letter to Mr. Gilbert, which is curious, as it shows that very nearly up to the time of the decomposition of the alkalies, his mind had been engaged on very different subjects.

TO DAVIES GIDDY, ESQ.

MY DEAR SIR, September 12, 1807.

I INCLOSE Mr. Carne's paper, which, when you have read, and Mr. Carne revised, I will thank you to inclose to me, and that as soon as possible, for the completion of the volume.

I have been a good deal engaged, since my return, in experiments on distillation, and I have succeeded in effecting what is considered of great importance in colonial commerce, namely, the depriving rum of its empyreumatic part, and converting it into pure spirit.

I mention this in confidence, as it is likely to be connected with some profitable results; and it may be beneficial in a public point of view, by lessening the consumption of malt.

I have heard of no scientific news; this, indeed, is little the season for active exertion.

With best respects to your father, and to Mr. and Mrs. Guillemard, I am, my dear Sir,

Always very faithfully yours,

H. DAVY.

Few notes have conveyed information of such importance to the scientific world, as that which follows, announcing, at the same time, the decomposition of the fixed alkalies, and the formation of the Geological Society, of which it would thus appear that Davy was one of the founders.

TO WILLIAM HASLEDINE PEPYS, ESQ.

DEAR PEPYS, November 13, 1807.

IF you and Allen had been one person, the Council of the Royal Society would have voted to you the Copleian Medal;* but it is an indivisible thing, and cannot be given to two.

We are forming a little talking Geological Dinner Club, of which I hope you will be a member. I shall propose you to-day. Some things have happened in the Chemical Club, which I think render it a less desirable meeting than usual, and I do not think you would find any gratification in being a member of it. Hatchett never comes, and we sometimes meet only two or three. I hope to see you soon.

I have decomposed and recomposed the fixed alkalies, and discovered their bases to be two new inflammable substances very like metals; but one of them lighter than ether, and infinitely combustible. So that there are two bodies decomposed, and two new elementary bodies found.

Most sincerely yours, H. DAVY.

* He alludes to a Paper, entitled "On the Quantity of Carbon in Carbonic Acid, and on the Nature of the Diamond; by William Allen, Esq. F.R.S. and William Hasledine Pepys, Esq." Communicated by Humphry Davy, Sec. R.S. M.R.I.A. Read June 16, 1807.

In the year 1808, MM. Gay Lussac and Thenard succeeded in decomposing potash by chemical means; for which purpose it is only necessary to heat iron turnings to whiteness in a curved gun-barrel, and then to bring melted potash slowly in contact with the turnings, air being excluded; when the iron, at that high temperature, will take the oxygen from the alkali, and the potassium may be collected in a cool part of the tube. It may likewise be produced by igniting potash with charcoal, as M. Curaudau showed in the same year.

In the following letter, Davy gives an account of his repeating the experiment of MM. Gay Lussac and Thenard; mixing together, as usual, science and angling.

TO J. G. CHILDREN, ESQ.

MY DEAR SIR, London, July 1808.

I HAVE this moment received your kind letter, and I have written to Pepys to propose to him to be with you on Sunday or Monday. I hope for his answer to-morrow morning, and I will write to you immediately.

I will procure all the fishing tackle you have proposed, and am most happy to find you in so determined a spirit for piscatory adventure.

I have had some letters from France; but nothing new, except an account of the gun-barrel experiment tolerably minute. I have tried it since, and procured potassium, but it was lost from some moisture passing into the aperture of the barrel. All that is necessary for the process is a gun-barrel bent thus, B ⁀⁀⁀⁀⁀ A.—B represents the part where the touch-hole is closed; here dry potash is introduced; and the middle, which is to be strongly ignited, contains the filings; the potash is gradually fused and made to run down upon the ignited iron; the potassium collects in A.

If you should be able to procure the apparatus for this experiment, I should like to assist in repeating it; and could we procure a large quantity of the basis, we may try its effects, on a great scale, on the undecompounded acids. I will bring some *dry boracic acid.* A copper or platina tube, if you have one, will be proper for trying the experiment in. We may likewise try its action upon the earths, and upon diamond.

I have metallized Ammonia,* without the application of Electricity. When an amalgam of potassium and mercury is brought in contact with an ammoniacal salt, the potassium seizes upon the oxygen, and the hydrogen and nitrogen unite to the quicksilver.

I had an opportunity of giving an account, on Friday, to the scientific men assembled at Greenwich, of your magnificent experiments and apparatus.† Sir Joseph Banks, Mr. Cavendish, Wol-

* He here alludes to a train of research, which will be considered hereafter.

† This observation relates to the magnificent battery constructed by Mr. Children, of which he presented an account to the Royal Society, in a Paper read in November 1808, entitled, "An Account of some Experiments, performed with a view to ascertain the most advantageous Method of constructing a Voltaic Apparatus, for the purpose of Chemical Research. By John George Children, Esq. F.R.S." The great battery described in this Paper consisted

laston, &c. all expressed a strong wish that the results should be published. I am most happy you have drawn up the account.

I regard the days I have passed in your society, as some of the pleasantest of my life. I look forward with a warm hope to our next meeting. Be pleased to assure your father of my highest respect, and of my gratitude for his kindness.

I am, my dear Sir,
Very sincerely yours,
H. DAVY.

It is impossible to reflect upon the chemical processes by which potassium may be obtained, without feeling surprised that the discovery should not have long before been accomplished. It is evident that the substance must have been repeatedly developed during the operations of chemistry; alkalies had been frequently heated to whiteness in contact both with iron and charcoal, and in some instances the appearance of a highly combustible body, which could have been no other than potassium, had even been observed as a result of the process, and yet no

of twenty pairs of plates, each plate being four feet high by two feet wide: the sum of all the surfaces was ninety-two thousand one hundred and sixty square inches, and the quantity of liquid necessary for charging it, one hundred and twenty gallons. At the same time he constructed another battery, which consisted of two hundred pairs of plates, each being only two inches square. In the one case, then, he commanded extent of surface, in the other, extent of number; and by a series of comparative experiments, he fully established the theory of Davy (page 246), that the *intensity* of Electricity increases with the *number*, and the *quantity* with the *surface*.

suspicion, as to its real nature, ever crossed the mind of the experimentalist; he satisfied himself with designating such a product, whenever it occurred, by the term *Pyrophorus.** I remember the late Mr. William Gregor informing me that, in the course of his analytical experiments with potash and different metals, he had repeatedly observed a combustion on removing the crucible from the furnace, and exposing its contents, which he could never understand. How admirably do such anecdotes illustrate the remark made in the commencement of the present chapter, that truth may be often touched, but is rarely caught, in the dark!

The facility of the combustion of the bases of the alkalies, and the readiness with which they decomposed water, offered Davy the ready means for determining the proportions of their constituent parts: and in comparing all his results, he thinks that it will be a good approximation to the truth, to consider potash as composed of about six parts base and one of oxygen; and soda, as consisting of seven base and two of oxygen.

The discovery of potassium led to that of the true nature of what had been long familiar to chemists by the name of *pure Potash,* but which ought to

* The Pyrophorus of Homberg, of which a description is to be found in the *Mémoires de l'Academie,* for 1711, was made by mixing together any combustible body, as gum, flour, sugar, charcoal, &c. and alum, and then, after roasting the mixture till it was reduced to a dry powder, exposing it in a matrass to a red heat. In this process, the theory of which was first explained by Davy, the potash of the alum is converted into potassium, which, by its absorption of oxygen from the atmosphere, generates heat, and sets fire to the charcoal contained in the powder.

have been called the *hydrat*, for the *pure* alkali was not known until after the discovery of Davy. The experiments of MM. Gay Lussac and Thenard have shown this substance to be a *Protoxide*. It is difficult of fusion; it has a grey colour and a vitreous fracture, and dissolves in water with much heat. The *Peroxide* is procured by the combustion of potassium at a low temperature; it had been observed by Davy in October 1807, but at that time he supposed it to be the oxide containing the smallest proportion of oxygen: it has a yellow colour, and when thrown into water effervesces, and gives out oxygen gas.* When heated very strongly upon platina, oxygen is also expelled from it, and there remains the *protoxide*, or pure potash.

It was a great object with Davy, to show that the product resulting from the combustion of potassium, was a pure oxide free from water; for it is evident that had potassium been a *Hydruret*, its combustion must have produced a *Hydrat*. This he accomplished by a series of experiments which he performed in the laboratory of Mr. Children, and which are published in his Bakerian Lecture of 1800.

Having discovered the presence of oxygen in the fixed alkalies, he was naturally led by analogy to enquire whether ammonia might not also contain it. It was true that the chemical composition of that body had been considered as satisfactorily settled, and that the conversion of it into hydrogen and

* The 'Potassa Fusa' of our Pharmacopœia generally contains a small proportion of the peroxide, and will therefore effervesce when thrown into water.

nitrogen, in the experiments of Scheele, Priestley, and Berthollet, had left nothing farther to be accomplished. All new facts, however, are necessarily accompanied by a new train of analogies ; and Davy, in perusing the accounts of the various experiments to which ammonia had been submitted, tells us that he saw no reason for considering the presence of oxygen as impossible ; for, supposing hydrogen and nitrogen to exist in combination with oxygen in low proportion, this latter principle might easily disappear in the analytical experiments by heat and electricity, in the form of water deposited upon the vessels employed, or dissolved in the gases produced.

Under this impression, he commenced a series of experiments by which, he says, he soon became satisfied of the existence of oxygen in the volatile alkali. By means of the Voltaic battery, he ignited perfectly dry charcoal in a small quantity of pure ammoniacal gas, and he produced carbonate of ammonia ; which could not have happened, had not oxygen been furnished by the volatile alkali to the carbon. In the next place, by an ingenious arrangement of apparatus, he submitted ammonia to a high temperature, and effected its decomposition, when a quantity of water appeared as one of the products. It will be useless to enter into farther details upon this occasion, as we shall presently perceive the subject assumed a different aspect, and led the experimentalist into a new line of enquiry.

At the conclusion of his Bakerian Lecture of 1807, he speaks of the probable composition of the earths, and considers it reasonable to expect that they are compounds of a similar nature to the fixed alkalies — "peculiar highly combustible metallic bases united to oxygen;"—but, as yet, this theory was sanctioned only by strong analogies; it was his good fortune, at a subsequent period, to support it by conclusive facts.

When the importance and novelty of the results he obtained from the fixed alkalies, and their influence upon the reigning theories, are duly considered, it may be easily imagined how intense was the curiosity to witness the production of the new metals, to examine their singular qualities, and to question the illustrious discoverer upon their nature and relations. Suppose it were publicly announced that one of the greatest Astronomers of the day had invented a telescope of new and extraordinary powers; and that by its means a hitherto unsuspected system of heavenly bodies might be seen, the character and motions of which were wholly inconsistent with the Newtonian theory of the universe. The surprise and eager curiosity which such an announcement would create might possibly be more general, because a knowledge of Astronomy is more widely diffused than that of Chemistry; but the sensation would not be more intense than that which the discovery of potassium produced. The laboratory of the Institution was crowded with persons of every rank and description; and Davy, as may be readily supposed, was kept in a continued state of excitement throughout the day. This circumstance, co-operating with the effects of the fatigue he had previously undergone, produced a most severe fit of illness, which for a time caused an awful pause in his researches, broke the thread of his pursuits, and turned his reflections into different channels.

He always laboured under the impression that the fever had been occasioned by contagion, to which he had been exposed in one of the jails during an experiment for fumigating it. This impression appears to have continued through life, for in his "Last Days" he alludes to it in terms of strong conviction.

Other persons referred his illness to the deleterious fumes, especially those of Baryta, to which his experiments had exposed him: an opinion recorded in an Epigram,* which was circulated amongst the members of the Institution after his recovery.

> Says Davy to Baryt, "I 've a strong inclination
> To try to effect your deoxidation;"
> But Baryt replies—"Have a care of your mirth,
> Lest I should retaliate, and change *you* to Earth."

Upon conversing, however, with Dr. Babington, who, with Dr. Frank, attended Davy throughout

* I received the above *jeu d'esprit* from the late learned Orientalist, Mr. Stephen Weston, only four days before his death. Since its publication in the first edition of this work, a chemical friend sent me the following improved version of it.

> Says Davy to Baryt, "I feel strong temptation
> To effect by my art your deoxidation;
> And the money I 've got in my pocket I 'll bet all,
> I prove you a true, though disguised lad of metal."
> Says Baryt to Davy, "A truce to your mirth;
> If you turn me to metal, I 'll turn you to earth;
> So moisten your clay, don't improve Science daily,
> Nor treat me as you 've treated poor Soda and Kali."

this illness, he assured me that there was not the slightest ground for either of these opinions; that the fever was evidently the effect of fatigue and an over-excited brain. The reader will not feel much hesitation in believing this statement, when he is made acquainted with the habits of Davy at this period. His intellectual exertions were of the most injurious kind, and yet, unlike the philosophers of old, he sought not to fortify himself by habits of temperance. Should any of my readers propose to me the same question respecting Davy, as Fontenelle tells us was put to an Englishman in Paris, by a scientific Marquis, with regard to Newton,—whether he ate, drank, and slept like other people? —I certainly should be bound to answer in the negative.

Such was his great celebrity at this period of his career, that persons of the highest rank contended for the honour of his company at dinner, and he did not possess sufficient resolution to resist the gratification thus afforded, although it generally happened that his pursuits in the laboratory were not suspended until the appointed dinner-hour had passed. On his return in the evening, he resumed his chemical labours, and commonly continued them till three or four o'clock in the morning; and yet the servants of the establishment not unfrequently found that he had risen before them. The greatest of all his wants was Time, and the expedients by which he economised it often placed him in very ridiculous positions, and gave rise to habits of the most eccentric description: driven to an extremity, he would in his haste put on fresh linen, without removing that which was underneath; and, singular as the fact may appear, he has been known, after the fashion of the grave-digger in Hamlet, to wear no less than five shirts, and as many pair of stockings, at the same time. Exclamations of surprise very frequently escaped from his friends at the rapid manner in which he increased and declined in corpulence.

At the commencement of his severe illness in November 1807, he was immediately attended by Dr. Babington and Dr. Frank; and upon its assuming a more serious aspect, these gentlemen were assisted by Dr. Baillie. Such was the alarming state of the patient, that for many weeks his physicians regularly visited him four times in the day, and issued bulletins for the information of the numerous enquirers who anxiously crowded the hall of the Institution. His kind and amiable qualities had secured the attachment of all the officers and servants of the establishment, and they eagerly anticipated every want his situation might require. The housekeeper, Mrs. Greenwood, watched over him with all the care and solicitude of a parent; and, with the exception of a single night, never retired to her bed for the period of eleven weeks. In the latter stage of his illness, he was reduced to the extreme of weakness, and his mind participated in the debility of the body. It may not perhaps be thought philosophical to deduce any inference, as to character, from traits which are displayed under such circumstances: I have little doubt, however, but that the mind, like inorganic matter, will, in its decay, frequently develope important elements, which under other conditions were not distinguishable. Suppose pride and timidity to exist as qualities in the same mind, the former might so far predominate as to enable its possessor to face the cannon's mouth; but diminish its force by moral or physical agency, the natural timidity will gain the ascendency, and the hero be converted into a coward. Such are my reasons for introducing the following anecdote: I would not give to it a greater value than it deserves, but it surely demonstrates the existence of kindly affections. Youthful reminiscences, and circumstances connected with his family and friends, were the only objects which, at this period, occupied his thoughts, and afforded him any pleasure. No Swiss peasant ever sighed more deeply for his native mountains, than did Davy for the scenes of his early years. He entreated his nurse to convey to his friends his ardent wish to obtain some apples from a particular tree which he had planted when a boy; and, unlike Locke with his cherries, he had no power of controlling the desire by his reason, but remained in a state of restlessness and impatience until their arrival: at the same time, he expressed a wish to obtain several other objects, especially an ancient tea-pot, endeared to him by early associations.

The following Minute appears on the Journals of the Institution:—

"December 7, 1807.—Mr. Davy having been confined to his bed for the last fortnight by a severe illness, the Managers are under the painful necessity of giving notice, that the Lectures will not commence until the first week in January next."

The Course was, at the time stated, opened by the Reverend Mr. (now Dr.) Dibdin; and his introductory lecture was, by order of the Managers, printed "for the satisfaction of those proprietors who were not present." It thus commences:— "Before I solicit your attention to the opening of those Lectures which I shall have the honour of delivering in the course of the season, permit me to trespass upon it for a few minutes, by stating the peculiar circumstances under which this Institution is again opened; and how it comes to pass that it has fallen to me, rather than to a more deserving lecturer, to be the first to address you.

"The Managers have requested me to impart to you that intelligence, which no one who is alive to the best feelings of human nature can hear without mixed emotions of sorrow and delight.

"Mr. Davy, whose frequent and powerful addresses from this place, supported by his ingenious experiments, have been so long and so well known to you, has for these last five weeks been struggling between life and death;—the effects of those experiments recently made in illustration of his splendid discoveries, added to consequent bodily weakness, brought on a fever so violent as to threaten the extinction of life. Over him, it might emphatically be said, in the language of our immortal Milton, that

———— 'Death his dart
Shook, but delayed to strike.'

Had it pleased Providence to have deprived the world of any further benefit from his original talents and immense application, there certainly has been already enough effected by him to entitle his name to a place amongst the brightest scientific luminaries of his country. That this may not appear an unfounded eulogium, I shall proceed, at the particular request of the Managers, to give you an outline of the splendid discoveries to which I have just alluded; and I do so with the greater pleasure, as that outline has been drawn in a very masterly manner by a gentleman of all others perhaps the best qualified to do it effectually."

The Lecturer then proceeded to take a general and rapid view of his labours, which it is unnecessary to introduce in this place, and concluded as follows:—

"This recital will be sufficient to convince those who have heard it of the celebrity which the author of such discoveries has a right to attach to himself; and yet no one, I am confident, has less inclination to challenge it. To us, and to every enlightened Englishman, it will be a matter of just congratulation, that the country which has produced the two BACONS and BOYLE, has in these latter days shown itself worthy of its former renown by the labours of CAVENDISH and DAVY. The illness of the latter, severe as it has been, is now abating, and we may reasonably hope that the period of convalescence is not very remote."

Fortunately for the cause of Science—fortunately for the interests of the Institution, the prediction of the learned Lecturer was shortly verified.

The Institution, indeed, had already suffered from the calamity; for, in a Report to the Visitors, dated January 25, 1808, it is stated, that "there has been an excess of expenditure beyond the receipts. Among the causes of diminished income may be mentioned the postponement of the lectures, in consequence of the lamented illness of the excellent Professor of Chemistry; and among the items of increased expenditure, the extra expense of the Laboratory, in which have been produced Mr. Davy's recent discoveries, so honourable to the Royal Institution, and so beneficial to the interests of science in every part of the world."

This Report is succeeded by the following Minute:—

"February 22, 1808.—Mr. Davy attended at the request of the Committee, and informed them that he should be able to commence his course of Lectures on Electro-chemical science on Saturday the 12th of March, at two o'clock; and those on Geology on Wednesday evening, the 16th of that month."

The following letter to Mr. Poole announces the restoration of his health, and communicates some other circumstances of interest. Mr. Poole, it would appear, entertained doubts as to whether Davy received the prize of France for his first or second Bakerian Lecture, upon which point this letter sets him right.

TO THOMAS POOLE, ESQ.

MY DEAR POOLE, March, 1808.

MANY thanks for your kind letter. I have seen your friend Mr. B—— for a minute, and, to use a geological term, I like his *aspect*, and shall endeavour to cultivate his acquaintance.

I am exceedingly busy; my health is re-established; and I am entering again into the career of experiment.

The prize which you congratulate me upon was given for my paper of 1806, and not for my last discoveries, which will probably excite more interest.

C——, after disappointing his audience twice from illness, is announced to lecture again this week. He has suffered greatly from excessive sensibility—the disease of genius. His mind is as a wilderness, in which the cedar and the oak, which might aspire to the skies, are stunted in their growth by underwood, thorns, briars, and parasitical plants. With the most exalted genius, enlarged views, sensitive heart, and enlightened mind, he will be the victim of want of order, precision, and regularity. I cannot think of him without experiencing the mingled feelings of admiration, regard, and pity.

Why do you not come to London? Many would be happy to see you; but no one more so than your very sincere friend, my dear Poole,

H. DAVY.

It is difficult to convey an adequate idea of the universal interest which was excited by the lectures on Electro-chemical Science, to which an allusion has been just made. The Theatre of the Institution overflowed; and each succeeding lecture increased the number of candidates for admission.

It is unnecessary, after what has been already stated, to describe the masterly style in which he demonstrated and explained those general laws which his genius had developed, or to enumerate the beautiful and diversified experiments by which he illustrated their application, in simplifying the more complex forms of matter.

His evening lectures on Geology were equally attractive; and by a method as novel as it was beautiful, he exhibited, by the aid of transparencies, the structure of mountains, the stratification of rocks, and the arrangements of mineral veins.

The Easter recess afforded him a few days of leisure, which from the following note he appears to have devoted to his favourite amusement.

TO W. H. PEPYS, ESQ.

MY DEAR PEPYS, April, 1808.

CHILDREN has had the kindness to arrange our party, and we are to meet him, at all events, on Tuesday at two o'clock, at Foot's Cray.

I have proposed that we should leave town at about five or six on Monday evening, sleep at Foot's Cray, and try the fly-fishing there.

Will you arrange with Allen, whom we must initiate in the vocation of the Apostles, as he wants nothing else to make him perfect as a primitive Christian and a Philosopher?

I am, my dear Pepys,
Most affectionately yours,
H. DAVY.

Hitherto his passion for angling has only been noticed in connection with his conversation and letters; I shall now present to the reader a sketch of the philosopher in his fishing costume. His whole suit consisted of green cloth; the coat having sundry pockets for holding the necessary tackle: his boots were made of caoutchouc, and, for the convenience of wading through the water, reached above the knees. His hat, originally intended for a coal-heaver, had been purchased from the manufacturer in its raw state, and dyed green by some pigment of his own composition; it was, moreover, studded with every variety of artificial fly which he could require for his diversion. Thus equipped, he thought, from the colour of his dress, that he was more likely to elude the observation of the fish. He looked not like an inhabitant o' the earth, and yet was on 't;—nor can I find any object in the regions of invention with which I could justly compare him, except perhaps with one of those grotesque personages who, in the farce of "The Critic," attend Father Thames on the stage, as his two banks.

I shall take this opportunity of stating, that his shooting attire was equally whimsical: if, as an angler, he adopted a dress for concealing his person, as a sportsman in woods and plantations, it was his object to devise means for exposing it; for he always entertained a singular dread lest he might be accidentally shot upon these occasions. When upon a visit to Mr. Dillwyn of Swansea, he accompanied his friend on a shooting excursion, in a broad-brimmed hat, the whole of which, with the exception of the brim, was covered with scarlet cloth.

Notwithstanding, however, the refinements which he displayed in his dress, and the scrupulous attention with which he observed all the minute details of the art; if the truth must be told, he was not more successful than his brother anglers; and here again the temperament of Wollaston presented a characteristic contrast to that of Davy. The former evinced the same patience and reserve — the same cautious observation and unwearied vigilance in this pursuit, as so eminently distinguished his chemical labours. The temperament of the latter was far too mercurial: the fish never seized the fly with sufficient avidity to fulfill his expectations, or to support that degree of excitement which was essential to his happiness, and he became either listless or angry, and consequently careless and unsuccessful. —But it is time to resume the thread of our chemical history.

It has been already stated, that Davy had no sooner decomposed the fixed alkalies than he proceeded to effect an analysis of the earths; but his results were indistinct: they could not, like the alkalies, be rendered conductors of electricity by fusion, nor could they be acted upon in solution, in consequence of the strong affinity possessed by their bases for oxygen. The pursuit of the enquiry then demanded more refined and complicated processes, than those which had succeeded with potash and soda.

The only methods which held out any fair prospect of success were those of operating by electricity upon the earths in some of their combinations, or of converting them, at the moment of their decomposition, into metallic alloys, so as to obtain presumptive evidence of their nature and properties. Such, in fact, was the line of enquiry in which Davy was deeply engaged, when he received from Professor Berzelius of Stockholm a letter, announcing the fact that he had, in conjunction with Dr. Pontin, succeeded in decomposing baryta and lime, by negatively electrising mercury in contact with them, and that, by such means, he had actually obtained amalgams of the earths in question.

Our philosopher immediately repeated the experiments, and with perfect success. After which he completed a series of additional experiments, which fully established the nature of these bodies, and the analogies he had anticipated. These results formed the subject of a memoir, which was read before the Royal Society on the 30th of June 1808, and entitled, "Electro-chemical Researches on the Decomposition of the Earths: with Observations on the Metals obtained from them, and on the Amalgam of Ammonia."

He commences this paper by enumerating the several trials he had made to effect the decomposition of these bodies; such as, First, by electrifying them by iron wires under the surface of naphtha, with a view to form alloys with iron and the metallic bases of the earths. Secondly, by heating potassium in contact with the alkaline earths, in the hope that this body might detach the oxygen from them, in the same manner as charcoal decomposes the common metallic oxides. Thirdly, by submitting various mixtures of the earths and potash to Voltaic action, with the idea that the potash and the earths might be deoxidated at the same time, and entering into combination, form alloys. Fourthly, by mixing together various earths with the oxides of tin, iron, lead, silver, and mercury: a mode of manipulation suggested by the results of his previous experiments on potassium, in which he found that when a mixture of potash and the oxides of mercury, tin, or lead, was electrified in the Voltaic circuit, the decomposition was very rapid, and an amalgam, or an alloy of potassium, was obtained; the attraction between the common metals and the potassium apparently accelerating the separation of the oxygen.

Supposing that a similar kind of action might assist the decomposition of the alkaline earths, he proceeded to institute a series of experiments upon that principle; and the results were more satisfactory than those obtained by the preceding methods of experimenting—a compound was obtained which acted upon water with the evolution of hydrogen, producing a solution of the earth, and leaving free the tin, or lead, with which its base may be supposed to have been alloyed;—but in all such experiments the quantity of the metallic basis produced must have been very minute, and its character very questionable.

In this stage of the enquiry, Davy received the letter from Professor Berzelius of Stockholm, the contents of which he embodied in his memoir, accompanied with such observations as his own information suggested.

" A globule of mercury, electrified by the power of a battery consisting of five hundred pairs of double plates of six inches square, weakly charged, was made to act upon a surface of slightly moistened barytes, fixed upon a plate of platina. The mercury gradually became less fluid, and after a few minutes was found covered with a white film of barytes; and when the amalgam was thrown into water, hydrogen was disengaged, the mercury remained free, and a solution of barytes was formed.

"The result with lime, as these gentlemen had stated, was precisely analogous.

"That the same happy methods must succeed with strontites and magnesia, it was not easy to doubt, and I quickly tried the experiment. From strontites I obtained a very rapid result; but from magnesia, in the first trials, no amalgam could be procured. By continuing the process, however, for a longer time, and keeping the earth continually moist, at last a combination of the bases with mercury was obtained, which slowly produced magnesia, by absorption of oxygen from air, or by the action of water.

"All these amalgams I found might be preserved for a considerable period under naphtha. In a length of time, however, they became covered with a white crust under this fluid. When exposed to air, a very few minutes only were required for the oxygenation of the bases of the earths. In the water the amalgam of barytes was most rapidly decomposed; that of strontites, and that of lime next in order: but the amalgam from magnesia, as might have been expected from the weak affinity of the earth for water, very slowly changed: when, however, a little sulphuric acid was added to the water, the evolution of hydrogen, and the production and solution of magnesia, were exceedingly rapid, and the mercury soon remained free."

In order, if possible, to procure the amalgams in quantities sufficient for distillation, he combined the methods he had employed in the first instance, with those pursued by Berzelius and Pontin. "A mixture of the earth with red oxide of mercury was placed on a plate of platina, a cavity was made in the upper part of it to receive a globule of mercury, the whole was covered by a film of naphtha, and the plate was made positive, and the mercury negative, by a proper communication with the battery of five hundred."

The amalgams thus procured were afterwards distilled in glass tubes filled with the vapour of naphtha; by which operation the mercury rose pure from the amalgam, and it was very easy to separate a part of it; but the difficulty was to obtain a complete decomposition, for to effect this, a high temperature was required, and at a red heat the bases of the earths instantly acted upon the glass, and became oxidated.

In the best result which Davy obtained in this manner, the barytic basis appeared as a white metal of the colour of silver, fixed at all common temperatures, but fluid at a heat below redness, and volatile at a heat above it. Unlike the alkaline bases, it would seem to be considerably heavier than water.

In extending these experiments to alumine, silex, zircone, &c. after a most elaborate investigation, such results were not obtained as justified the conclusion that they were, like the other earths, metallic oxides; although, as far as they went, they added to the probability of such analogy.

It will be remembered that, after the fixed alkalies had been found to contain oxygen, Davy was very naturally led to enquire whether ammonia might not also contain the same element, or be an oxide with a binary base. In the communication from Professor Berzelius, and Dr. Pontin, already alluded to, a most curious experiment is related on what they consider the deoxidation and amalgamation of the compound basis of ammonia; and which they regard as supporting the idea which Davy had formed of the presence of oxygen * in the volatile alkali. A fact so startling as the production of a metallic body from ammonia, or from its elements, immediately excited in Davy's mind the most ardent desire to pursue the enquiry; and, after repeating the original experiments of the Swedish chemists with his accustomed sagacity, he modified his methods of manipulation, in order, if possible, to obtain this metallic body in its most simple form; but, although he succeeded in producing the amalgam without Voltaic aid, by the intervention of potassium, he could not so distill off the mercury as to leave the basis, or imaginary *ammonium*, free.

The history of these researches into the nature of the ammoniacal element concludes the lecture of which I have endeavoured to give an outline. The subject of the amalgam is still involved in mystery: if we suppose with Davy, that a substance, which forms so perfect an amalgam with mercury, must of necessity be metallic in its own nature, we cannot but conclude either that hydrogen and nitrogen are both metals in the aëriform state, at the usual temperatures of the atmosphere—bodies, for example, of the same character as zinc and quicksilver would be at the heat of ignition—or, that these gases are

* See page 285.

oxides in their common form, but which become metallized by deoxidation—or, that they are simple bodies, not metallic in their own nature, but capable of composing a metal in their deoxygenated, and an alkali in their oxygenated, state.

Before we venture, however, to entertain any opinions so extravagant in their nature, and so wholly unsupported by analogy, it would be well to enquire how far the change, which ammonia and mercury undergo by Voltaic action, really merits the name of amalgamation. Several chemists of the present day are inclined to refer this change of form to a purely mechanical cause, by which the particles of the metal become separated, and converted, as it were, into a kind of *froth* by the operation.*

* The late interesting experiments of Mr. Daniell "on the action of Mercury on different Metals," which have been recently published in the first number of a new series of the Journal of the Royal Institution, appear to throw much light upon this subject. By agitating a few grains of spongy platinum with mercury in water acidified with acetic acid, he obtained an amalgam of the consistence of soft butter, which retained its consistence for many weeks, and greatly resembled that formed by the electrization of mercury in contact with ammonia. When the amalgam was laid upon filtering paper, the moisture was gradually absorbed and evaporated, and the mercury returned to a fluid state. By a more refined experiment, Mr. Daniell ascertained that the process was accompanied by the evolution of hydrogen gas; whence he very fairly concludes, that, when minutely divided, platinum is agitated with mercury, and moisture is present, an electrical action takes place, which, when heightened by the addition of a diluted acid, or the solution of a neutral salt, is sufficiently energetic to decompose water and evolve hydrogen: the oxygen at the same time combines with the mercury, and a solution is effected by the acetic acid, which its unassisted affinity could not have produced. It also appears," continues Mr. Daniell, "that this electrical action commu-

Mr. Brande, in a late communication in the Journal of the Royal Institution,* observes: " Shortly after the discovery of a method of obtaining Morphia in a pure state, I remember that Sir Humphry Davy suggested the possibility of its affording,

nicates an adhesive attraction to the particles of the metal, by which the particles of liquid and aëriform bodies are entangled and retained, a kind of *frothy* compound formed, and the fluidity of the mercury destroyed. The appearance of this amalgam is so very like that of the ammoniacal compound formed by exposing a solution of ammonia in contact with mercury to the influence of the Voltaic pile, or when an amalgam of potassium and mercury is placed upon moistened muriate of ammonia, that it is impossible not to be struck with the resemblance. Mr. Daniell is therefore inclined to believe that the production of the latter may be explained upon the same principles as that of the former. When the effect is produced by the direct application of the electrical current, by means of the battery, it ceases the moment the connexion between the poles is broken; and when brought about by the agency of the amalgam of potassium, the electrical action is doubtless excited by the contact of the two dissimilar metals, and the frothy compound lasts no longer than the existence of the potassium in the metallic state; whereas in the action between mercury and finely divided platinum, the permanence of the metals produces a much more lasting effect, and the soft amalgam may therefore be preserved for a greater length of time.

* " On the Electro-chemical Decomposition of the Vegeto-Alkaline Salts." In this communication, the Professor gives an account of some experiments of his own, with a view to ascertain whether the vegetable alkalies, if electrised in contact with mercury, would impart any principle to the latter metal. In experiments with morphia and cinchonia, in which the mercury in contact with the vegetable base was rendered negative, not the least change in the fluidity of the metal could be perceived. When, however, a similar experiment was made with quina, the metal became filmy, and acquired even a tendency to a butyraceous appearance, but the phenomenon was found to depend upon the presence of a minute portion of lime.

when electrised in contact with mercury, results corresponding with those which Berzelius had observed in respect to ammonia. He thought that the nascent elements of the morphia, as liberated by electrical decomposition, might, under such circumstances, effect a similar apparent amalgam of the mercury, and he spoke of the subject as likely to throw some light upon the corresponding ammoniacal combinations. He made, I believe, a few experiments upon the subject; but as the results were not such as he had anticipated, they were not placed on record."

In the progress of our ascent, it is refreshing to pause occasionally, and to cast a glance at the horizon, which widens at every increase of our elevation. By the decomposition of the alkalies and earths, what an immense stride has been made in the investigation of nature!—In sciences kindred to chemistry, the knowledge of the composition of these bodies, and the analogies arising from it, have opened new views, and led to the solution of many problems. In Geology, for instance, has it not shown that agents may have operated in the formation of rocks and earths, which had not previously been known to exist? It is evident that the metals of the earths cannot remain at the surface of our globe; but it is probable that they may constitute a part of its interior; and such an assumption would at once offer a plausible theory in explanation of the phenomena of volcanoes, the formation of lavas, and the excitement and effects of subterranean heat, and might even lead to a general theory in Geology.

The reader, for the present, must be satisfied with

these cursory hints: I shall hereafter show that our illustrious philosopher followed them up by numerous observations and original experiments in a volcanic country.

I remember with delight the beautiful illustration of his theory, as exhibited in an artificial volcano constructed in the theatre of the Royal Institution.—A mountain had been modelled in clay, and a quantity of the metallic bases introduced into its interior: on water being poured upon it, the metals were soon thrown into violent action—successive explosions followed—red-hot lava was seen flowing down its sides, from a crater in miniature—mimic lightnings played around: and in the instant of dramatic illusion, the tumultuous applause and continued cheering of the audience might almost have been regarded as the shouts of the alarmed fugitives of Herculaneum or Pompeii.

CHAPTER VIII.

Davy's Bakerian Lecture of 1808.—Results obtained from the mutual action of Potassium and Ammonia upon each other.—His belief that he had decomposed Nitrogen.—He discovers Telluretted Hydrogen.—Whether Sulphur, Phosphorus, and Carbon, may not contain Hydrogen.—He decomposes Boracic acid.—Boron.—His fallacies with regard to the composition of Muriatic acid.—A splendid Voltaic Battery is constructed at the Institution by subscription.—Davy ascertains the true nature of the Muriatic and Oxy-muriatic Acids.—Important chemical analogies to which the discovery gave origin.—Euchlorine.—Chlorides.—He delivers Lectures before the Dublin Society.—He receives the Honorary Degree of LL.D. from the Provost and Fellows of Trinity College.—He undertakes to ventilate the House of Lords.—The Regent confers upon him the honour of Knighthood.—He delivers his farêwell Lecture.—Engages in a Gunpowder manufactory.—His marriage.

THE third Bakerian Lecture, which Davy read before the Royal Society in December 1808, is entitled "An Account of some new analytical Researches on the Nature of certain Bodies, particularly the Alkalies, Phosphorus, Sulphur, Carbonaceous matter, and the Acids hitherto undecompounded; with some general Observations on Chemical Theory."

The object of this lecture was to communicate the results of numerous experiments which had been instituted for the purpose of still farther extending our knowledge of the elements of matter, by the new powers and methods arising from the application of electricity to chemical analysis.

Important as were the facts thus obtained, they disappointed the expectation of those who did not consider, that the more nearly we approach ultimate analysis,* the greater must be the difficulties, the more numerous the fallacies, and the less perfect the results, of our processes. In fact, his former discoveries had spoilt us: their splendour had left our organs of perception incapable of receiving just impressions from any minor lights, and we participated with exaggerated feelings, in the disappointment which he himself expressed at several of his results. The confidence inspired by his former

* The difficulty of seizing upon elementary forms, as well as the infinity of combinations of which they are susceptible, are supposed by Mr. Sankey to be allegorized in the fable of Proteus, Πρωτειος, being derived from πρωτος, signifying the first element. It is not a little singular that Mr. Leslie, to whom such a speculation was wholly unknown, should have recognised in the same fable a picture of the cautious but intrepid advances of the skilful experimenter: he tries to press Nature into a corner,—he endeavours to separate the different principles of action,—he seeks to concentrate the predominant agent, and labours to exclude as much as possible every disturbing influence. Notwithstanding the confidence with which modern philosophers have claimed the discovery, the experimental mode of investigation was undoubtedly known and pursued by the ancients, who appear, observes Mr. Leslie, to have concealed their notions respecting it under the veil of allegory. *Proteus* signified the mutable and changing forms of material objects; and the inquisitive philosopher was counselled by the Poets to watch their slippery demon, when slumbering on the shore to bind him, and to compel the reluctant captive to reveal his secrets.—*Elements of Natural Philosophy.*

triumphs may be compared to that which is felt by an army, when commanded by a victorious General, — a conviction that, however difficult may be the enterprise, it must be accomplished by the genius of him who undertakes it. The moment we discovered that Davy was laying siege to one of Nature's strongest holds,—that he was attempting to resolve nitrogen into other elementary forms,—we regarded the deed as already accomplished, and the repulse which followed most unreasonably produced a feeling of dissatisfaction. Upon such occasions, the severity of our disappointment will always be in proportion to the importance of the object we desire to accomplish; and it is impossible not to feel that the discovery of the true nature of nitrogen would lead to new views in chemistry, the extent of which it is not easy even to imagine.

The principal objects of research which this paper embraces are,—the elementary matter of ammonia; the nature of phosphorus, sulphur, charcoal, and the diamond; and the constituents of the boracic, fluoric, and muriatic acids. Enquiries which are continued and extended in two successive papers, *viz.* in one read before the Society in February 1809, entitled "New Analytical Researches on the Nature of certain Bodies; being an Appendix to his Bakerian Lecture of 1808;" and in his fourth Bakerian Lecture of 1809, "On some new Electro-chemical Researches on various Objects, particularly the Metallic bodies from the Alkalies and Earths; and on some Combinations of Hydrogen."

With regard to these admirable papers,—for such they must undoubtedly be considered,—the biographer must confine his observations to their general character and results. They are far too refined to admit of a brief analysis, and too elaborate to allow a successful abridgement. A just idea of their merit can alone be derived from a direct reference to the Philosophical Transactions.

The enquiry commences with experiments on the results produced by the mutual action of potassium and ammonia on each other. His object was two-fold: to refute the hypothesis which assumed hydrogen as an element of potassium, and to ascertain the nature of the matter existing in the amalgam of ammonia, or the supposed metallic basis of the volatile alkali: a question intimately connected with the whole of the arrangements of chemistry. As to the former point, it is unnecessary to enter into farther discussion; and with regard to the latter, it is quite impossible to convey an adequate idea of the extent of the enquiry: there does not exist in the annals of chemistry a more striking example of experimental industry.

In the course of his experiments on potassium and ammonia, he obtained an olive-coloured body, which he was inclined to regard as a compound of the metallic base of ammonia (*ammonium*) and potassium; and on submitting which to various trials, he uniformly obtained, as the product of its decomposition, a proportion of nitrogen considerably less than that which, upon calculations founded on a rigid analysis of the volatile alkali, ought to have been afforded under such circumstances, while the potassium employed at the same time became oxidated. This result inspired a hope that nitrogen might have been actually decomposed during the process, and that its elements were oxygen and a metallic basis, or oxygen and hydrogen.

That he was sanguine in that hope, appears from the whole tenor of his paper; in farther proof of which, I can adduce a letter which he addressed to Mr. Children during the progress of his experiments, in which he says, "I hope on Thursday to show you nitrogen as a complete wreck, torn to pieces in different ways." His subsequent enquiries, however, although they did not strengthen the suspicion he had formed respecting the decomposition of that body, yet indirectly developed facts of considerable importance; which, with his characteristic quickness of perception, he made subservient to fresh investigation.

His researches into the phenomena exhibited by tellurium, when forming a part of the Voltaic circuit, are highly interesting. It had been stated by Ritter, that, of all the metallic substances he tried for producing potassium by negative electricity, tellurium was the only one by which he could not procure it; and he uses this fact in support of his opinion, that potassium is a *hydruret.* He says, that when a circuit of electricity is completed in water by means of two surfaces of tellurium, oxygen is given off at the positive surface, and instead of hydrogen at the negative surface, a brown powder is formed and separated, which he regards as a *hydruret* of tellurium; and he conceives that the reason why that metal prevents the metallization of potash is, that it has a stronger attraction for hydrogen than that possessed by the alkali.

Davy's attention was naturally arrested by such a statement, and, in pursuing the enquiry, he discovered a series of new facts:—he found that tellurium and hydrogen were capable of combining, and of forming a gas, to which he gave the name of *telluretted hydrogen,*—that, so far from tellurium preventing the decomposition of potash, it formed an alloy with potassium when negatively electrified upon the alkali—and, such was the intense affinity of potassium and tellurium for each other, that the decomposition of potash might be effected by acting on the oxide of the latter metal and the alkali, at the same time, by heated charcoal.

With respect to the next subject of enquiry in these papers, *viz.* whether sulphur, phosphorus, and carbon, in their ordinary forms, may not contain hydrogen, it would appear that from an experiment performed by Mr. Clayfield, and which Davy witnessed at Bristol in the year 1799, he was very early led to suspect the existence of hydrogen in sulphur; but it was not until 1807, that he entered upon the investigation of the subject. From the general tenor of his experiments he concluded that, in its common state, it may be regarded as a compound of small quantities of oxygen and hydrogen, with a large quantity of a basis which, on account of its strong attractions for other bodies, has not hitherto been obtained in its pure form. The same analogies apply to phosphorus and carbon. His conclusion was mainly derived from the fact, that hydrogen is produced from sulphur and phosphorus in such quantities by Voltaic electricity, that he thinks it cannot well be considered as an accidental ingredient in them: the presence of oxygen, he contends, may be inferred from the circumstance that, when potassium is made to act upon these bodies, the sulphurets and phosphurets so formed evolve by the action of an acid less hydrogen, in the form of compound inflammable gas, than the same quantity of potassium in an uncombined state. The question, however, still remains in considerable doubt; and in his "Elements of Chemical Philosophy," published four years afterwards, he admits that no accurate conclusions have been formed on the subject.

In his second Bakerian Lecture of 1807, Davy had given an account of an experiment in which boracic acid appeared to be decomposed by Voltaic electricity, a dark-coloured inflammable substance separating from it on the negative surface. In the memoir now under consideration, he procured the basis by heating together boracic acid and potassium, when he ascertained it to be a peculiar inflammable matter, which, after various experiments upon its nature, he was inclined to regard as metallic; on which account he proposed for it the name of *Boracium*. At about the same period, MM. Gay Lussac and Thénard were engaged in investigating the same subject in France, and they anticipated him in some of the results.

When Davy, by subsequent experiments, had ascertained that the base of the boracic acid is more analogous to carbon than to any other substance, he adopted the term *Boron*, as less exceptionable than that of *Boracium*.

At this time, he also entered upon the investigation of fluoric acid, the results of which must be reserved for future consideration.

His experiments and reasonings upon muriatic acid, at this period of his career, must be now considered as deriving their greatest degree of interest from their fallacy; and they deserve an examination in this work, if it be only to estimate the vigour he subsequently displayed in disentangling himself from a web of his own fabrication. The most satisfactory proof of intellectual strength is to be found in the existence of a power which enables the mind to conquer its prejudices and to correct its own errors. How many remarkable instances does the history of science present, in which the philosopher has treated his facts as Procrustes did his victims, in order that they might accord with the measure most convenient for his purpose!

Prejudiced by the general opinion respecting the hitherto undecompounded nature of muriatic acid, he had long sought to discover its radical by the agency of Voltaic electricity; but he uniformly found that when its aqueous solution was thus acted upon, the water alone underwent decomposition; while the electrization of the gas afforded no other indication of its nature than the presence of a much greater quantity of water than theory had assigned to it. He proceeded, therefore, to examine the acid by other modes of enquiry: he found, by the action of potassium upon the gas, that a large volume of hydrogen was evolved, which, in conjunction with other experiments, satisfied him that this body, in its common aëriform state, contained at least one-third of its weight of water; and he adopted various expedients with the hopes of obtaining the acid free from it. Without pursuing him through this research, I shall merely state the conclusions at which he arrived, *viz.* that dry muriatic acid, could it be obtained, would probably be found to possess the strongest and most extensive powers of combination of all known substances belonging to the class of acids; and that its basis, should it ever be separated in a pure form, will be one of the most powerful agents in Chemistry. From the fact of water appearing in a separate state, and oxymuriatic acid being formed whenever a metallic oxide was heated in muriatic acid gas, he was led to consider the muriatic acid as a compound of a certain base, (not hitherto obtained in a separate state,) and not less than one-third part of water; while he regarded oxymuriatic acid as a compound of the same base (free from water) with oxygen.

After the numerous experiments in which the original battery of the Institution had been used, so greatly were its metallic plates corroded, that it was found to be no longer serviceable; in consequence of which, as it would appear from a minute, dated July 11, 1808, "Mr. Davy laid before the Managers of the Royal Institution the following paper, *viz.*

"A new path of discovery having been opened in the agencies of the electrical battery of Volta, which promises to lead to the greatest improvements in Chemistry and Natural Philosophy, and the useful arts connected with them; and since the increase of the size of the apparatus is absolutely necessary for pursuing it to its full extent, it is proposed to raise a fund by subscription, for constructing a powerful battery, worthy of a national establishment, and capable of promoting the great objects of science.

"Already, in other countries, public and ample means have been provided for pursuing these investigations. They have had their origin in this country; and it would be dishonourable to a nation so great, so powerful, and so rich, if, from the want of pecuniary resources, they should be completed abroad.

"An appeal to enlightened individuals on this subject can scarcely be made in vain. It is proposed that the instrument and apparatus be erected in the Laboratory of the Royal Institution, where it shall be employed in the advancement of this new department of science."

The Minute goes on then to state that—

"The above paper having been laid before the Board of Managers, they felt it their indispensable duty instantly to communicate the same to every member of the Royal Institution, lest the slightest delay might furnish an opportunity to other countries for accomplishing this great work, which originated in the brilliant discoveries recently made at the Royal Institution.

"The Managers present agree to subscribe to this undertaking.

"ORDERED, that a book be opened at the Steward's office for the purpose of entering the names of all those members who may wish to contribute towards this important National object."

To the great gratification of Davy, and to the honour of the country, the list of subscribers was soon completed, and one of the most magnificent batteries ever constructed was speedily in full operation.

It is thus alluded to in his Elements of Chemical Philosophy :—"The most powerful combination that exists, in which number of alternations is combined with extent of surface, is that constructed by the subscriptions of a few zealous cultivators and patrons of science, in the Laboratory of the Royal Institution. It consists of two hundred instruments connected together in regular order, each composed of ten double plates arranged in cells of porcelain, and containing in each plate thirty-two square inches; so that the whole number of double plates is two thousand, and the whole surface one hundred and twenty-eight thousand square inches."

This battery, when the cells were filled with sixty parts of water mixed with one part of nitric acid, afforded a series of brilliant and impressive effects. When pieces of charcoal, about an inch long, and one-sixth of an inch in diameter, were brought near each other, (within the thirtieth or fortieth parts of an inch,) a bright spark was produced, and more than half the volume of the charcoal became ignited to whiteness, and by withdrawing the points from each other a constant discharge took place through the heated air, in a space equal at least to four inches, producing a most brilliant ascending arch of light, broad and conical in form in the middle. When any substance was introduced into this arch, it instantly became ignited; platina melted as readily in it as wax in the flame of a common candle; quartz, the sapphire, magnesia, lime, all entered into fusion; fragments of diamond, and points of charcoal and plumbago rapidly disappeared, and seemed to evaporate in it, even when the connexion was made in a receiver exhausted by the air-pump; but there was no evidence of their having previously undergone fusion.

All the phenomena of chemical decomposition were produced with intense rapidity by this combination. When the points of charcoal were brought near each other in non-conducting fluids, such as oils, ether, and oxymuriatic compounds, brilliant sparks occurred, and elastic matter was rapidly generated.

Among the numerous experiments performed by the aid of this battery, he instituted several, in the hope of decomposing nitrogen; and which are recorded in his Bakerian Lecture of 1809. He ignited potassium, by intense Voltaic electricity, in this gas; and the result was, that hydrogen appeared, and some nitrogen was found deficient. This, on first view, led him to the suspicion that he had attained his object; but, in subsequent experiments, in proportion as the potassium was more free from a coating of potash, which necessarily introduced water, so in proportion was less hydrogen evolved, and less nitrogen found deficient. The general tenor of these enquiries, therefore, did not strengthen the opinion he had formed with respect to the compound nature of nitrogen.

It appears from the following letter, that Davy

visited his friend Mr. Andrew Knight at Downton, in September 1809. It is introduced in these memoirs principally for the purpose of showing with what boldness he was accustomed to depart from generally received opinions, and to project new theories for the explanation of the most abstruse subjects.

TO JOHN GEORGE CHILDREN, ESQ.

MY DEAR FRIEND, September 23, 1809.

I AM about to visit Downton, and shall return by the first of October. I have neither seen nor heard from Lord Darnley, and I conjecture he has not yet returned from Scotland.

I wish you great sport in pheasant-shooting, but I trust you have had still nobler game in your Laboratory.

I doubt not you have found before this, as I have done, that the substance we mistook for *sulphuretted* hydrogen is *telluretted* hydrogen, very soluble in water, combinable with alkalies and earths, and a substance affording another proof that hydrogen is an oxide. I have met with another analogous compound, that of *boracium* with hydrogen, which possesses very similar properties.

I find that taking *ammonium* as the basis of hydrogen, according to the ideas which I stated, all the compounds will agree with the suppositions that I mentioned to you, *viz.* eight cubic inches of hydrogen, two of oxygen, ammonia; four and two, water; four and four, nitrogen; four and six, nitrous oxide; four and eight, nitrous gas; four and ten, nitric acid. Where the multiples are not in geometrical order, the decomposition is most easy, *i. e.* in nitrous oxide and nitric acid; more easy in water than in ammonia; but most difficult in nitrogen, where there is probably the most perfect equilibrium of affinities.

I have kept charcoal white hot by the Voltaic apparatus, in dry oxymuriatic acid gas for an hour, without effecting its decomposition. This agrees with what I had before observed with a red heat. It is as difficult to decompose as nitrogen, except when all its elements can be made to enter into new combinations.

I find the radiation, *in vacuo,* from ignited platina, is to that in air as three to one:—so much for Leslie's hypothesis.

A little electrical machine acts with a repulsion as two, in a vacuum equal to five inches of mercury; as thirty, in common air; as thirty, in oxygen; as twenty-nine or thirty, in hydrogen; and as forty-five, in carbonic acid. I showed this experiment, made with every precaution, to Mr. Cavendish, Dr. Herschell, Dr. Wollaston, and Warburton: so much for the theory, that electricity is dependent upon oxidation. I do not think our worthy friend Pepys will resist any longer.

Pray let me know what you have been doing. I hope you will not suffer these beautiful and satisfactory experiments of the capacities of metals to remain still. Write me a letter as egotistical as the one I have given you. You are pledged to do good and noble things, and you must not disappoint the men of science of this country.

With kindest remembrances to your excellent

father, and with hopes that we shall soon meet, I am, my dear friend,

Very faithfully and affectionately yours,

H. DAVY.

The genius displayed by Mr. Knight in investigating the phenomena of vegetable nature, and in applying the knowledge thus acquired to objects of practical improvement, excited in Davy, as might have been expected, feelings of the highest admiration; and when, in addition to such claims, he was the acknowledged patron and hospitable friend of the angler, the reader will readily imagine the warmth of feeling with which our philosopher cherished his friendship.

On commencing the present work, I applied to Mr. Knight for any assistance he might be able to afford me, in aid of so arduous a labour; and he very kindly returned an answer, from which I extract the following passage.

"My late lamented friend, Sir Humphry Davy, usually paid me a visit in the autumn, when he chiefly amused himself with angling for grayling, a fish which he appeared to take great pleasure in catching. He seemed to enjoy the repose and comparative solitude of this place, where he met but few persons, except those of my own family, for we usually saw but little company. He always assured me that he passed his visits agreeably, and I had reason to believe he expressed his real feelings.

"In the familiar conversations of these friendly visits, he always appeared to me to be a much more extraordinary being than even his writings, and vast discoveries, would have led me to suppose him; and, in the extent of intellectual powers, I shall ever think that he lived and died without an equal."

The reader has already been made acquainted with those experiments which led Davy to modify the prevailing opinions, with regard to the constitution of the muriatic and oxymuriatic acids; and on the false assumption that oxygen existed in the latter gas, to refer the deposition of water which takes place upon heating a metallic oxide in the former, to the supposition that muriatic acid contains a large proportion of water as essential to its composition. Upon observing, however, that charcoal, if freed from hydrogen and moisture, even when ignited to whiteness in oxymuriatic, or muriatic acid gas, by the Voltaic battery, did not effect the least change in them, he was led to suspect the accuracy of his previous conclusion; and on retracing his steps, and entering upon a new path of enquiry, he ultimately succeeded, after one of the most acute controversies that ever sprang from a chemical question, in recalling philosophers to the original theory of Scheele, by establishing the important truth, that oxymuriatic acid is, in the true logic of chemistry, a simple body, which becomes muriatic acid by its union with hydrogen.

The new views arising out of such a revolution in chemical opinion are certainly not the least important of those to which the discoveries of Davy have given birth. Dr. Johnson has remarked, that "one of the most hazardous attempts of criticism is to choose the best amongst many good." I am much mistaken, however, if the chemists of Europe will not, without hesitation, pronounce his researches into the nature of oxymuriatic acid, and its relations, with the exception of those by which he established the chemical laws of Voltaic action, to be by far the most important of all his labours; not only as evincing the ascendancy of his genius, and the steadiness of his perseverance, but as marking a new and splendid era in chemical science.

It is much more difficult to eradicate an ancient error than to establish a new truth; and on this occasion, he had not only to contend against the pampered errors of a domineering system, but against the equivocal and illusive evidence, or, if I may be allowed the expression, the apparent neutrality of facts by which the truth of his theory was to be judged. In consequence of the constant and often unsuspected interference of water, there is scarcely a result connected with the chemical history of the bodies in dispute, that did not admit of being equally well explained upon the hypothesis that oxymuriatic acid is a compound, as upon that of its being a simple or undecompounded substance. The question could never have been determined but by an investigation of the most refined and subtile nature; so delicately was the evidence balanced, that nothing but the keenest eye, and the steadiest hand, could have determined the side on which the beam preponderated.

The illustrious discoverer of oxymuriatic acid considered that body as muriatic acid freed from hydrogen, or, in the obscure language of the Stahlian school, as muriatic acid deprived of phlogiston, whence he assigned to it the name of *dephlogisticated* muriatic acid. Upon the establishment of the anti-phlogistic theory by Lavoisier, it became essential to the generalization which distinguished it, that a body performing the functions of an acid, and above all, supporting the process of combustion, should be regarded as containing oxygen in its composition; and facts were not wanting to sanction such an inference. The substance could not even be produced from muriatic acid, without the action of some body known to contain oxygen; while the fact of such a body becoming deoxidated by the process, seemed to demonstrate, beyond the possibility of error, that the conversion of the muriatic into the oxymuriatic acid, was nothing more than a simple transference of oxygen from the oxide to the acid: an opinion which was universally adopted, and which for nearly thirty years triumphed without opposition.

The body of evidence by which Davy overthrew this doctrine, and established the undecompounded nature of oxymuriatic acid, is to be found in a succession of papers read before the Royal Society, *viz.* in that already announced,—in his Bakerian Lecture for 1810,—and in a subsequent memoir read in February 1811.

It will be impossible for me to follow the author through all the intricacies of the enquiry; but I shall seize upon some of its more prominent points, and give a general outline of its bearings.

No sooner had his suspicions been excited with regard to the compound nature of oxymuriatic acid, than it occurred to him that, if oxygen were really present in that body, he might readily obtain it from some of its compounds; that, for instance, its combination with tin would yield an oxide of that metal by ammonia; while those with phosphorus would furnish, on analysis, either the phosphor*ous*, or phosphor*ic* acid. But after experiments in which the presence of water was most cautiously excluded, the results he had anticipated were not obtained. In the place of an oxide of tin, the product, on the application of heat, volatilized in dense and pungent fumes; and, instead of obtaining an acid of phosphorus, a body possessing new and unexpected properties resulted. Again,—it had been stated, in confirmation of the theory that recognised the presence of oxygen in oxymuriatic acid, that when this latter body and ammonia were made to act upon each other, water was formed: our chemist frequently repeated the experiment, and convinced himself that such was not the fact.

It had been shown by Mr. Cruickshank, and more recently proved by MM. Gay Lussac and Thénard, that oxymuriatic acid and hydrogen, when mixed in nearly equal proportions, produce a matter almost entirely condensable by water, which is common muriatic acid; and that water is not deposited in the operation. Davy made many experiments on the subject, and he found, that when these gases were mingled together in equal volumes over water, introduced into an exhausted vessel, and fired by the electric spark, muriatic acid resulted, although, at the same time, there was a certain degree of condensation, and a slight deposition of vapour; but on repeating the experiment in a manner still more refined, and by carefully drying the gases, such condensation became proportionally less.

When, in addition to the above experimental evidence, it is stated that MM. Gay Lussac and Thénard had proved, by a copious collection of instances, that in the usual cases where oxygen is eliminated from oxymuriatic acid, water is always present, and muriatic acid gas is formed; and as it has been moreover shown that oxymuriatic is converted into muriatic acid gas by combining with hydrogen, it is scarcely possible to avoid the conclusion, that the oxygen is derived from the decomposition of water, and not from that of the acid.

When mercury is made to act, by means of Voltaic electricity, upon one volume of muriatic acid gas, all the acid disappears, calomel is formed, and half a volume of hydrogen is evolved.

By such experiments and arguments, Davy was led to the conclusion that, as yet, oxymuriatic acid has not been decompounded; that it is a peculiar body, elementary as far as our knowledge extends, and analogous, in its tendency of combination with inflammable matter, to oxygen gas; that, in fact, it may be a *peculiar* acidifying and dissolving principle, forming with different substances compounds analogous to acids containing oxygen, or to oxides, in their properties and powers of combination, but differing from them in being, for the most part, decomposable by water. On this idea, he thinks that muriatic acid may be considered as having hydrogen for its base, and oxymuriatic acid for its acidifying principle. In confirmation of such an opinion, it is also important to remark, that in its electrical relations, oxymuriatic acid maintains its analogy to oxygen.

The vivid combustion of bodies in oxymuriatic acid gas, Davy acknowledges, may, at first view, appear a reason why oxygen should be admitted as one of its elements; but he answers this argument by stating, that heat and light are merely results of the intense agency of combination; and that sulphur and metals, alkaline earths and acids, become alike ignited under such circumstances.

As change of theory with regard to the primitive must necessarily modify all our views with respect to the nature of secondary bodies, so must this new view of oxymuriatic acid affect all our opinions respecting its compounds. Davy accordingly proceeded, in the first place, to investigate the various bodies which had been distinguished by the name of *hyper-oxymuriates, muriates,* &c.

It also became necessary to alter the nomenclature, since to call a body which neither contains oxygen nor muriatic acid, by a term which denotes the presence of both, is contrary to those very principles which first suggested it. Having consulted some of the most eminent philosophers, Davy proposed a name founded upon one of the most obvious and characteristic properties of the oxymuriatic acid, namely, its colour, and called it CHLORINE.

If then oxymuriatic acid, or chlorine, does not contain any oxygen, a question immediately arises as to the true nature of those compounds in which the muriatic acid has been supposed to exist in combination with a much larger proportion of oxygen than in the oxymuriatic acid,—in the state in which it has been named by Mr. Chenevix *hyper-oxygenized* muriatic acid.

In his Bakerian Lecture of 1810, entitled, "On some of the Combinations of Oxymuriatic Gas and Oxygen, and on the Chemical Relations of these Principles," he details a number of experiments for the illustration of this subject, and arrives at the conclusion, that the oxygen in the hyper-oxymuriate of potash is in triple combination with the metal and chlorine. He likewise confirms his views, with regard to the elementary nature of this latter body, by a series of new enquiries, and shows that they are not incompatible with known phenomena:—for instance, Scheele explained the bleaching powers of oxymuriatic gas, by supposing that it destroyed colours by combining with *Phlogiston.* Berthollet* considered it as acting by imparting oxygen; Davy now proves that the pure gas is wholly incapable of altering vegetable colours, and that its operation in bleaching entirely depends upon its property of decomposing water, and of thus liberating its oxygen.† The experiment by which he demonstrated this fact is so simple and satisfactory, that I shall here relate it. Having filled a glass globe, containing dry powdered muriate of lime, with oxymuriatic gas, he introduced into another globe, also containing muriate of lime, some dry paper tinged with litmus,

* Berthollet first applied oxymuriatic acid for the purpose of bleaching, in France; from whence Mr. Watt introduced it into England.

† Dr. Thomson has more recently explained the operation, by supposing that water is decomposed, and that its hydrogen goes to the chlorine, and its oxygen to the water, forming with the latter a deutoxide of hydrogen, or the oxygenated water of Thénard, which he considers as the true bleaching principle.

that had been just heated; by which device the intrusion of moisture was effectually prevented. After some time, this latter globe was exhausted, and then connected with that containing the oxymuriatic gas, and by an appropriate set of stopcocks, the paper was exposed to the action of the gas thus dried: no change of colour in the test paper took place, and after two days, there was scarcely a perceptible alteration; while some similar paper dried and introduced into the gas, that had not been exposed to muriate of lime, was instantly bleached.

As an illustration of the eagerness with which he seized upon facts, in order to apply them to economical purposes, it may be stated that, on reflecting upon the theory of bleaching, and on the changes which its agents undergo, he was led to propose the use of a liquor produced by the condensation of oxymuriatic gas in water, containing magnesia diffused through it, as superior to the oxymuriate of lime commonly employed.*

It has been very truly observed, that all knowledge which is gained tends towards the acquisition of more, just as the iron dug from the mine facilitates in return the working of the miner. Never

* Experience has not confirmed the value of this suggestion. Davy imagined that the vegetable fibre was injured by the saline residuum; and having found that muriate of magnesia was less corrosive than muriate of lime, he was led to propose the substitute above stated. The fact, however, is, that the fibre is injured by the chlorine; and as this body has only a slight affinity for magnesia, it too quickly abandons it; and consequently the oxymuriate of lime is still preferred.

was this truth more forcibly illustrated than by the discovery of the nature of chlorine. In the progress of that train of enquiry, which became necessary for the adjustment of our views as they regarded the combinations of that body, Davy discovered a series of new compounds, the history of which he communicated in successive papers to the Royal Society.

In a memoir read in February 1811, entitled, "On a Combination of Oxymuriatic Gas and Oxygen Gas," he announced the existence of a *protoxide* of chlorine, under the name of *Euchlorine;* and in a communication from Rome in the year 1815, he described another compound of chlorine and oxygen, containing a still larger proportion of this latter element, and which has since been made the subject of a series of experiments by Count Stadion of Vienna. As it does not exhibit any acid properties, Dr. Henry proposes to call it a *Peroxide,* in preference to *Deutoxide;* thinking it probable that intermediate compounds, between this and the protoxide already mentioned, may be hereafter discovered.

His paper on *euchlorine* abounds with interest. He found that by acting on the salts formerly denominated *hyper-oxymuriates,* by muriatic acid, the gas evolved differed very greatly in its properties, with the different modes of preparing it. When much acid was employed with a small quantity of the salt, and the gas was collected over water, it was not found to differ from oxymuriatic gas; but when, on the other hand, the gas was procured by means of a weak acid, and a considerable excess of the salt, at a low heat, and was collected over mercury, it possessed properties essentially different. Its colour, under such circumstances, was of a dense tint of brilliant yellow-green, whence the name of *euchlorine.** When in a pure form, this gas is so readily decomposed, that it will sometimes explode during the time of its transfer from one vessel to another, producing both heat and light with an expansion of volume,† and it may always be made to explode by a very gentle heat, often even by that of the hand.

The results of its explosion indicate its composition to be one atom of chlorine, and one of oxygen. None of the metals that burn in chlorine act upon this gas at common temperatures; but when the oxygen is separated, they then inflame in the residual chlorine. This fact Davy illustrated by a series of experiments, one of which, from its extreme beauty, I shall here relate. If a glass vessel, containing copper-leaf, be exhausted, and the euchlorine afterwards admitted, no action will take place; but throw in a little nitrous gas, and a rapid decomposition will ensue, and the metal will burn with its accustomed brilliancy.

The discovery of this interesting gas, and that of the facts connected with it, not only confirmed the novel views with regard to the elementary na-

* From ευ and χλωρος.

† The most vivid effects of combustion known are those produced by the *condensation* of oxygen, or chlorine: but in this instance, a violent explosion with heat and light is produced by their separation, and *expansion;* a perfectly novel circumstance in chemical philosophy.

ture of chlorine, but they reconciled the contradictory accounts of different authors respecting the properties of that body.

The weak attraction subsisting between the elements of this compound gas, which by a comparatively low temperature are made repulsive of each other, confirms also the supposition of Davy, that oxygen and chlorine belong to the same class of bodies.

The discovery of the *peroxide of chlorine* was made during an examination of the action of acids on the *hyper-oxymuriates* of Chenevix, undertaken by Davy in consequence of a statement of M. Gay Lussac, that a peculiar acid, which he called *chloric acid,* might be procured from the *hyper-oxymuriate of baryta* by sulphuric acid. With regard to this acid, which its discoverer considered as composed of one atom of chlorine and five atoms of oxygen, Davy entered into a warm controversy, affirming that the fluid in question owed its acid powers to combined hydrogen; and that it was analogous to the other hyper-oxymuriates, as being triple compounds of inflammable bases with chlorine and oxygen, in which the two former determine the character of the compound: this opinion, however, he afterwards abandoned, and I have reason to believe that he regretted ever having advanced it.

Amidst these new views, it became necessary to alter our opinions with regard to many of those compounds which have been termed *muriates,* but which, it would appear, contain neither muriatic acid nor oxygen, but are, strictly speaking, combinations of metals with chlorine, held in union by a very powerful affinity, since chlorine is capable of expelling the whole of the oxygen from any metallic oxide, and of taking its place; even those metals that are most distinguished by their affinity for oxygen, abandon it whenever their oxides are heated in chlorine, in which case oxygen gas is disengaged.

The same metal is also capable of uniting with different proportions of chlorine, which, so far as has been yet ascertained, are definite, and in no case exceed two proportions to one of metal. Hence it was proposed by Davy, in fixing the nomenclature of these compounds, to designate such as contain the least proportion of chlorine by the termination *ane*, added to the Latin name of the metal, as *cuprane* for that of copper; those containing the larger proportion of chlorine, by the termination *anea*, as *cupranea*. The chemical name of our common culinary salt, in conformity with such a nomenclature, would be *sodane*. This proposition, however, has not been adopted;* the compounds of metals and chlorine are either called *chlorurets*, or what is preferable, from their analogy with the similar compounds of oxygen, *chlorides*, and which are further distinguished as *protochlorides*, *deutochlorides*, &c.

In connexion with the history of these chlorides, a question arises of great interest and obscurity, and

* A little reflection will convince us that such a nomenclature could never have been adopted with propriety. It is in direct defiance of the Linnæan precept, that a specific name must not be united to the generic as a termination; besides which, such terms could never have been preserved in translations into other languages.

which has engaged the attention of some of our most distinguished chemists,—whether such a body, when dissolved by water, remains as a chloride; or, by decomposing that fluid, and combining with its elements, is not immediately converted into a muriate? With respect to several of these chlorides, no doubt can be entertained as to the fact of their decomposing water; for instance, the chloride of phosphorus is thus acted upon, the oxygen of the water forms phosphorous acid with the phosphorus, while its hydrogen unites with the chlorine to form muriatic acid; and as those products are such as do not combine with each other, but exist in a state of mixture in the water, each may be recognised by its peculiar properties. In like manner, as Davy has observed, when water is added in certain quantities to Libavius's liquor (*deutochloride of tin*), a solid crystalline mass is obtained, from which oxide of tin and muriate of ammonia can be obtained by ammonia.

In his Elements of Chemical Philosophy, Davy has been, in many instances, explicit on this point; and his opinions are favourable to the idea that chlorides become muriates by being dissolved in water: thus, he states that the perchloride of iron "acts with violence upon water, and forms a solution of red muriate of iron;" and he observes that the permuriate "forms a solution of green muriate of iron by its action upon water."* With regard,

* For an admirable paper upon this subject by Mr. R. Phillips, in which all the material points of the subject are considered with that acumen which distinguishes its author, see Annals of Philosophy, vol. i. New Series.

however, to the general principle, that chlorides become muriates by solution, there are difficulties which do not fall within the province of a biographer to discuss. I shall merely observe that such a change is, in many cases, so inconsistent with our preconceived opinions, that very strong evidence is required to reconcile us to its truth. We are undoubtedly prepared to hear that much may happen between the cup and the lip,—but that common salt should be a *chloride of sodium* on our plates, and a *muriate of soda* in our mouths, is certainly a very startling assertion.

The reception which the chloridic theory met with from the chemical world might aptly enough be adduced in illustration of that remark with which I commenced the preceding chapter. At first, its truth was questioned, and no sooner had this been triumphantly established, than an attempt was invidiously made to transfer the glory of the discovery from Davy to the French philosophers. Upon each of these points, I shall beg to offer a few observations.

First, with regard to the fact of chlorine being as yet an undecompounded body. The very announcement of a theory so adverse to the universal faith of Europe, was a signal for open hostilities; the observations of Dr. Murray may be considered as expressing the sentiments of most of the leading chemists on the first publication of the novel views of Davy. "Opinions," says he, "more unexpected have seldom been announced to chemists, than those lately advanced by Mr. Davy with regard to the constitution of the muriatic and oxymuriatic acids;

viz. that the latter is not a compound of muriatic acid and oxygen, but a simple substance, and that the former is a compound of this substance with hydrogen. The more general principle connected with these opinions, that oxymuriatic acid is, like oxygen, an acidifying element, forming with inflammables and metals an extensive series of analogous compounds, leads still more directly to the subversion of the established chemical systems, and to an entire revolution in some of the most important doctrines of the science."

Dr. Murray entered the lists as the avowed partisan of the theory of Berthollet; Dr. Davy, on the other hand, appeared as the champion of his brother's doctrine. A severe contest ensued, and both combatants displayed equal skill and strength. The object of the former was to demonstrate the presence of water, or its elements, as a constituent part of muriatic acid; and he proposed to determine the point by combining the dry gases of muriatic acid and ammonia; for as these bodies did not contain its elements, should water appear, he maintained that it must be considered as pre-existing in the muriatic acid; while, on the contrary, if no water could be procured, it would be unphilosophical to suppose it present, but that muriatic acid gas must, in that case, be considered as a compound of hydrogen and chlorine. In performing this experiment, Dr. Murray did succeed in obtaining a portion of water; but the inference from such a fact was questioned on the other side, upon the assumption of the humidity of the gases. As all parties, however, seemed to agree, that if every source of error could be excluded, the combination of these gases would furnish an *experimentum crucis*, by which the truth or fallacy of either theory might be established, Davy, when at Edinburgh, was desirous of repeating the experiment with Dr. Hope, and it was accordingly made in the College Laboratory. Sir George Mackenzie, Mr. Playfair, and some other gentlemen, were present. The results were communicated in Nicholson's Journal by Dr. Davy, and may be briefly stated as follows :—The alkaline and acid gases were pure, and both had been previously dried by exposure for sixteen hours to substances having a strong attraction for water. The apparatus consisted of a plain retort of about the capacity of twenty-six cubic inch measures, with a stop-cock; and of a receiver, with a suitable stop-cock. The latter was filled over mercury with one of the gases, which from the receiver passed into the exhausted retort by means of the stop-cocks; the other gas was introduced the same way into the retort; and thus alternately about ninety cubic inches of each gas were combined. All the salt having then been driven into the bulb of the retort by the heat of a spirit lamp, the neck was cooled and kept cold by moistened cloths, whilst the bulb was heated by a coke fire, till the muriate began to sublime, and to make its appearance at the curvature of the vessel when the fire was withdrawn. The result was then examined, while the bottom of the retort was still very hot: a dew, just perceptible, was observed lining the cold neck. The quantity of water was so extremely small, that the globular particles composing this dew could scarcely be perceived by the naked eye; now the quantity of water, according to hypothesis, should equal no less than eight grains. There is no small difference, it must be confessed, between that quantity and a dew barely perceptible, and which may reasonably be referred to a minute quantity of vapour in the gases, or to a little moisture derived from the mercury, a small quantity of which entered the retort with the gases. Dr. Hope wished to ascertain how much water would produce such a dew as was observed. For this purpose he heated in a retort, of a similar size to that used in the experiment, a single drop of water, which it may be said weighs about a grain. The appearance of condensed water, in this instance, in the neck of the retort, was much greater than in the preceding: he considered it as being three or four times as great.*

From these results it may be concluded, on Dr. Murray's own ground of reasoning, that water is not a constituent part of muriatic acid gas, and that this substance is a compound merely of chlorine and hydrogen; for it is easy to account for the presence of about one-third of a grain of water from various sources, while it is impossible to account for the absence of eight grains upon any theory except that which supposes the gas to be *anhydrous*.

I shall not pursue the numerous other experiments by which it was attempted to prove the fallacy of Davy's views; they all turn upon the same point, and were refuted by the same vigorous me-

* Sir Humphry Davy, during some experiments on the diamond, subsequently ascertained that less than 100th of a grain of water is sufficient to produce a sensible dew on a polished surface.

thods of enquiry. The chloridic theory may therefore now be considered as fully established: the philosophers who were for so long a period hostile to its reception, have at length yielded their assent; and Berzelius, in a paper published in the "Annales de Chimie," on the subject of sulpho-cyanic acid, has unconditionally tendered his allegiance; while the subsequent discovery of iodine and bromine has confirmed, by the most beautiful analogies, the views so satisfactorily explained by experiment.

As to the claim of priority which has been urged by several philosophers in favour of the French chemists, Davy, in speaking of Gay Lussac's paper, published in the "Annales de Chimie" for July 1814, observes, that "the historical notes attached to it are of a nature not to be passed over without animadversion. M. Gay Lussac states, that he and M. Thénard were the first to advance the hypothesis that chlorine was a simple body; and he quotes M. Ampère as having entertained that opinion before me. On the subject of the originality of the idea of chlorine being a simple body, I have always vindicated the claims of Scheele; but I must assume for myself the labour of having demonstrated its properties and combinations, and of having explained the chemical phenomena it produces; and I am in possession of a letter from M. Ampère, that shows he has no claims of this kind to make."*

The question of priority appears to me to be readily settled by a reference to printed documents. Davy published his "Elements of Chemical Philo-

* Royal Institution Journal, vol. i. p. 283.

sophy" in 1812, containing a systematic account of his new doctrines concerning the combinations of simple bodies. Chlorine is there placed in the same rank with oxygen, and finally removed from the class of acids. In 1813, M. Thénard published the first volume of his "Traité de Chimie Elémentaire Théorique et Pratique," in which he states the composition of oxymuriatic acid as follows:—"*Composition.* The oxygenated muriatic gas contains the half of its volume of oxygen gas, not including that which we may suppose in muriatic acid." It was not until the year 1816, that, by a note in his fourth volume, he appears to have at all relaxed in his attachment to the old theory of Lavoisier and Berthollet; and it will presently appear, that at the period above mentioned, iodine had been discovered, and its analogies to chlorine fully established, by the sagacity of Davy.

Having, as I trust, offered an impartial view of his claims to the establishment of the chloridic theory, I shall resume my narrative of those events which more immediately connect themselves with his personal history at this period.

The great fame of Davy, and the high importance of the discoveries which had bestowed it, became a general theme of admiration throughout the scientific circles of Europe, and induced the members of the Dublin Society to invite him to that city, for the purpose of delivering a course of lectures. From the authentic documents which have been placed in my hands, I am enabled to give a particular account of this transaction.

At a meeting of the Dublin Society held on the

3rd of May 1810, the following Resolutions were proposed and unanimously carried, *viz.*

1. "That it is the wish of the Society to communicate to the Irish public, in the most extended manner consistent with the other engagements of the Society, the knowledge of a Science so intimately connected with the improvement of Agriculture and the Arts, which it is their great object to promote; and that, with this view, it appears to them extremely desirable to obtain the fullest information respecting the recent discoveries made by Mr. Davy, in Electro-chemical science.

2. "Resolved, That application be made to the Royal Society, requesting that they will be pleased to dispense with the engagements of Mr. Davy, so far as to allow the Dublin Society to solicit the favour of his delivering a course of Electro-chemical Lectures in their new Laboratory, as soon as may be convenient after their present course of chemical lectures shall have been completed by their Professor Mr. Higgins.

3. "That the sum of four hundred guineas be appropriated out of the funds of the Society, to be presented to Mr. Davy, as a remuneration for the trouble and expense which they propose he should incur, and as a mark of the importance they attach to the communication which they solicit."

Mr. Leslie Foster having stated to the Dublin Society that the "Farming Society of Ireland" were desirous of availing themselves of this opportunity to apply to Mr. Davy to repeat before them the six lectures on the application of chemistry to agriculture, which he delivered this year (1810) to the

Board of Agriculture in England, and that they requested the Dublin Society would accommodate them with the use of their Laboratory for that purpose, all the members of the Dublin Society having free admission to such lectures—

The following Resolutions were passed by the Dublin Society:—

"That in the event of Mr. Davy coming over to Ireland, and consenting to deliver the Course referred to, the Farming Society shall be accommodated with the use of the Laboratory, according to their request.

"That it be referred to the Committee of Economy to consider on what terms, and under what regulations, it may be expedient to issue tickets of admission to the Electro-chemical Course, so as to reimburse to the Society the expenses attendant on the arrangement; and that, in order to give the fullest effect to such regulations, the members of the Society renounce any claim to gratuitous admission to this course."

A letter having been addressed to Mr. Davy by the Secretary of the Society, inviting him to Dublin, for the purpose of delivering courses of lectures, in conformity with the foregoing resolutions, the following answer was received from him:—

TO JOHN LESLIE FOSTER, ESQ. M. P. SECRETARY TO THE DUBLIN SOCIETY.

SIR, May 30, 1810.

I HAD the honour of communicating your letter to the President and Council of the Royal Society, who have desired me to express to you, Sir, and through you, to the Dublin Society, the lively interest they feel in the prosperity of that useful public body, and the desire that they have to promote its important object.

On these grounds, they have been pleased to permit me to be absent from the meetings of the Royal Society, during the period that may be necessary for delivering a Course of Lectures at the Laboratory of the Dublin Society, in the month of November next.

Be pleased to express to the Dublin Society my grateful acknowledgments for the honour they have done me in making such a proposition; and assure them that I shall use my best exertions to promote their views for the extension of Chemical Science, and every other species of useful knowledge.

I beg to be permitted to thank you, Sir, for the flattering manner in which you had the goodness to convey to me their proposal.

I am, Sir, with great respect,

Your obliged and obedient servant,

H. DAVY, SEC. R. S.

On the commencement of the Course, on the 8th of November 1810, three hundred and seventy-one admission tickets had been issued; and the Committee of Chemistry having expressed their opinion to the Society, that the lecture-room would not afford accommodation for a greater number of persons, the Assistant Secretary was directed to limit his tickets to that number. On the 15th instant, however, the number was increased to four hundred, without inconvenience.

At the close of the Course, on the 29th of November, the Dublin Society passed the following Resolutions:—

"Resolved, That the thanks of the Society be communicated to Mr. Davy for the excellent Course of Lectures which, at their request, has been delivered by him in their Laboratory; and to assure him that the views which led the Society to seek for these communications, have been answered even beyond their hopes;—that the manner in which he has unfolded his discoveries has not merely imparted new and valuable information, but further appears to have given a direction of the public mind towards Chemical and Philosophical enquiries, which cannot fail in its consequences to produce the improvement of the Sciences, Arts, and Manufactures in Ireland.

"That the thanks of the Society be communicated to the Royal Society for their ready compliance with our request, in dispensing with the engagements of Mr. Davy, during the last six weeks.

"That Mr. Davy be requested to accept the sum of five hundred guineas from the Society."*

The following letter appears, from the date, to have been written about a week before his arrival in Dublin.

TO THOMAS POOLE, ESQ.

MY DEAR POOLE, October 12, 1810.

Upon every occasion your recommendation, or opinion, would have great weight with me.

* There were four hundred tickets issued for the Course, sixty of which were honorary; the produce of the remainder amounted to 672*l.* 5*s.* 3*d.* Davy received 525*l.*; and the surplus went to officers and servants, and for the discharge of incidental expenses.

Amongst the candidates for the office of Clerk to the Royal Society, there is one Mr. W——, that I am well acquainted with, and who was formerly attached to the Royal Institution. He appears to me, as well from his scientific character, as from his habits and pursuits, to be admirably fitted for the situation. I advised him nearly two months ago, in consequence of a conversation with Sir Joseph Banks, to offer himself for the situation. I cannot therefore interest myself for any other person who does not possess superior qualifications.

Sir Joseph's maxim, which I hope will be adopted by all the members, is—"let it be given to the most worthy." I have no doubt that Mr. —— would fill the situation with credit, and that he is a very worthy man; but, from all that I can learn, his claims are much inferior to those of W——. We want not merely a civil, gentlemanlike, honest man, but a man a little accustomed to calculation, to astronomical observation, and to experiment.

I am in a delightful country here—the Valley of the Tyne—enjoying a few days' leisure after a rather hard chemical campaign, and preparing health and spirits for another in Ireland, where I am going next week.

I hope to be in London by the first week in December. I intend next summer to go into Cornwall —God willing; and I will not go through without seeing you, and telling you that, under all circumstances, I shall always think of you with the warmest esteem, and shall always be

Your sincere friend,

H. DAVY.

In the following year, Davy was again solicited by the Dublin Society to deliver lectures in their laboratory; and at a meeting of the members on the 13th of June 1811, a series of resolutions were passed, by which he was empowered to procure copies of many of the geological sketches referred to in a course of lectures he had delivered on Geology at the Royal Institution; and also to superintend the construction of a large Voltaic battery, for the illustration of the proposed lectures.

In compliance with this request, Davy delivered two distinct courses; one on the Elements of Chemical Philosophy, the other on Geology, for which he received the unanimous thanks of the Society, and as a more substantial testimony of their gratitude, the sum* of seven hundred and fifty pounds; the receipt of which Davy acknowledged by the following letter.

TO B. MAC CARTHY, ESQ. ASSISTANT SECRETARY TO THE DUBLIN SOCIETY.

SIR, Dublin, December 9, 1811.

I HAVE received your letter, inclosing a draught for seven hundred and fifty pounds Irish.

I am very much gratified by the thanks of the Dublin Society, for the courses of lectures which I had the honour of delivering in their laboratory; and I am proud of their opinion, that they will be useful to the Irish public.

* These Courses were more numerously attended than those in 1810, there having been issued about five hundred and twenty-five tickets; the proceeds of which were 1101*l.* 2*s.*

The attention, candour, and indulgence with which they were received by the audience, I shall remember with the warmest feelings of gratitude as long as I live.

I have the honour to be, Sir, with much esteem, your obliged and obedient servant, H. DAVY.

Before he quitted Dublin, the Provost and Fellows of Trinity College conferred upon him the honorary degree of LL.D., as an expression of the high admiration which his eminent scientific merits had so universally commanded.

In the month of August, in the same year, his opinion was requested by a committee, as to the best method to be adopted for ventilating the House of Lords; to which circumstance he alludes in the following note to his friend Mr. Pepys.

MY DEAR PEPYS, August 10, 1811.

I FIND that I am engaged on Wednesday, to meet Lord Liverpool, at the House of Lords, to consider a mode of ventilating it.

This business, most unluckily, will prevent my accompanying you; but I shall be glad to go with you on some other day, and to touch up the trout at Cheynies, and afterwards to proceed to Serge Hill.

Very affectionately yours, H. DAVY.

This undertaking, it must be allowed, was on Davy's part a most complete failure: whether he had miscalculated the diameter and number of the apertures necessary for establishing a current, it is difficult to say, but it was obvious that the stream of fresh air thus introduced was by no means adequate to the demand for it. *

The failure, so vexatious to Davy, became to others a fertile source of pleasantry, and numerous epigrams, not exactly of a character to meet the public eye, were very generally circulated, and which, in recording the miscarriage of science, displayed the triumph of wit.

The scientific renown of Davy having attracted the attention of his late Majesty, at that time Prince Regent, he received from his Royal Highness the honour of Knighthood, at a levee held at Carlton House, on Wednesday, the 8th of April 1812; and it may be remarked, that he was the first person on whom that honour had been conferred by the Regent.

On the day following this occurrence, Sir Humphry delivered his farewell lecture before the members of the Royal Institution; for he was on the eve of assuming a new station in society, which induced him to retire from those public situations which he had long held with so much advantage to the world, and with so much honour to himself. How far such a measure was calculated to increase his happiness I shall not enquire; but I am bound to observe,

* In February 1812, he exhibited a model, in one of his lectures at the Royal Institution, in illustration of his plan; from which it appeared that the air deteriorated by respiration was conducted through three copper pipes, terminating in a single tube, to the roof of the building; and by means of ventilators below, there was a constant supply of fresh air, the circulation of which was promoted by a furnace.

that it was not connected with any desire to abandon the pursuit of science, nor even to relax in his accustomed exertions to promote its interests. It was evident, however, to his friends, that other views of ambition than those presented by achievements in science, had opened upon his mind: the wealth he was about to command might extend the sphere of his usefulness, and exalt him in the scale of society: his feelings became more aristocratic, he discovered charms in rank which had before escaped him, and he no longer viewed Patrician distinction with philosophic indifference.

On the 11th of April 1812, Sir Humphry married Mrs. Apreece, the widow of Shuckburgh Ashby Apreece, Esq. eldest son of Sir Thomas Apreece: this lady was the daughter and heiress of Charles Kerr, of Kelso, Esq. and possessed a very considerable fortune.

Immediately after the celebration of the marriage, Sir Humphry and his bride proceeded to the hospitable mansion of Sir John Sebright, and afterwards made a tour through Scotland, receiving wherever they went the most flattering marks of attention.

During their excursion, Davy wrote various letters to his scientific friends, several of which I shall introduce; but, in order that those to Mr. Children may be understood, it will be necessary that the reader should be made acquainted with a transaction which occurred in the year 1811.

In consequence of some conversation on gunpowder, during which Davy observed that its composition might be greatly improved by rendering it less *hygrometric*, a proposition was started, that he should join Mr. Children and Mr. Burton in establishing a manufactory for its preparation upon chemical principles. Whether Davy considered himself, in the strict commercial sense, a partner, or merely a chemical adviser, it is perhaps not easy to determine; but it is quite clear that both Mr. Children and Mr. Burton considered him in the former light, although it is an act of justice to those gentlemen to state, that the very moment Davy expressed his disinclination to such an arrangement, they immediately, without the slightest hesitation, released him from all responsibility. This I am enabled to assert, after a most careful investigation of all the correspondence that passed upon the occasion.

TO JOHN GEORGE CHILDREN, ESQ.

Harewood House, July 14, 1812.

MY DEAR FRIEND,

I AM very sorry that I missed you the day before I set out on my journey. You will have learnt from your solicitor that I signed the articles. I still think I shall return before any powder will be made, at least if you do not make it till December, for our present intention is to be in town early in that month.

I sent to you an imperfect copy of my book,* in which there were no engravings, and in which one cancel was not inserted, thinking that you would prefer a copy sent in that way: the cancelled leaf, which you have not, contains a correction for the quantity of nitrous acid gas and water

* "Elements of Chemical Philosophy," to be presently noticed.

to form the crystalline compound, which is the base of oil of vitriol. Three parts nitrous acid gas condense four parts sulphurous gas.

I have my little apparatus, which will enable me to pursue my experiments on gunpowder. There is one conclusion very obvious resulting from the new facts,—a *perfect* gunpowder ought to contain no more charcoal than is necessary to convert the oxygen of the nitre into carbonic acid. Sulphur forms from nitre just as much elastic fluid as charcoal, *i. e.* if similar quantities of nitre be entirely decomposed, one by charcoal, and one by sulphur, and if the sulphurous gas and the carbonic acid gas be compared, their volumes will be equal. The advantage of forming carbonic acid gas is, that it is more readily disengaged from the alkali. Now it is a question, whether sulphur will decompose *sulphate* of potash,—it will decompose the carbonate; of this we are sure.

There ought, then, to be just as much sulphur as will form sulphuret of potash with the potash: 191 of nitre, 28·5 of charcoal, and 30 of sulphur, are the true proportions for forming nothing but sulphuret of potash and elastic matter.

Pray send me some cards to circulate; address to me, Post Office, Edinburgh. I hope you got Cavendish's balance.

I have been here for two days:—it is a very magnificent place: good fishing for pike, trout, and grayling. Lady D. desires her kind remembrances.

I am, my dear friend,

Most affectionately yours,

H. DAVY.

TO THE SAME.

Dunrobin Castle, near Golspie, August 21.

MY DEAR FRIEND,

I HOPE you are making progress in our manufactory. I shall expect, on my return, to find your powder the best and strongest, and to make trial of it. I wish I had some of it here, the black-cock and grouse would feel its efficacy. I have been expecting a letter from you every day.

This house is so delightful, the scenery so grand, and the field-sports so perfect, that I think we shall not quit it for a fortnight.

I went to Inverness and fished for salmon. I also went to two or three other places, but not one did I catch till I arrived here. The first day I landed seven noble ones, and played three more in four or five hours. The next day I played eight and landed three, besides white trout in abundance.

I have shot only one day, for a few hours; but we found grouse at every fifty yards, and I shot seven. We are just going to try sea-fishing.

Pray write to me a little news of what is doing for science and the world.

I beg you will remember me most kindly to your father and to Dr. Babington, and Brande, when you see them.

I am, my dear friend,

Most affectionately yours,

H. DAVY.

TO WILLIAM CLAYFIELD, ESQ.

Dunrobin, near Golspie, August 28, 1812.

DEAR CLAYFIELD,

I AM much obliged to you for two very kind letters, and for a box containing specimens from St. Vincent.* I beg you will thank the gentleman who was so good as to cause them to be collected for me. The box followed me to Inverness.

The ashes, I think, are likely to fertilize Barbadoes. There is a parallel case of materials having been carried so far in the eruption in Iceland in 1783.

I have been with my wife making a tour through the North since the beginning of July. We have arrived at our extreme point, and shall slowly proceed South in about a fortnight.

I wish you could be of our party here; we are in a delightful house, that of Lord Stafford, in a country abounding with fish and game. I have caught about thirty salmon since I have been here, and killed grouse, wild ducks, teal, &c. I have not yet shot a stag, but I hope to do so this next week.

I have just published a volume of the Elements of Chemistry, and I hope to publish another in the course of the Spring.

Having given up lecturing, I shall be able to devote my whole time to the pursuit of discovery.

* Specimens of substances ejected from the crater in that island, which Mr. Clayfield forwarded to Davy, in consequence of having heard that he had been engaged in examining the sand collected at Barbadoes, and which was a product of the same eruption.

I have not sent you a copy of my book, for I have thought that the best mode of avoiding giving offence to some, was by not making presents at all. Had I not so determined, one of the first copies would have been sent to you, as a mark of the warm esteem and regard of

Your affectionate friend, H. DAVY.

TO SAMUEL PURKIS, ESQ.

MY DEAR PURKIS, Dunrobin Castle, Aug. 29, 1812.

YOU may probably be surprised to receive a letter from me from this remote corner of the North; but I owe you a letter, and I have a great inclination, wherever I may be, to discharge all debts, and particularly those rendered due by kindness.

Receive my warm acknowledgments for your kind congratulations on my becoming a Benedick. I can now speak from experience, in which you have long participated. I am convinced that the natural state of domestic society is the best fitted for man, whether he be devoted to philosophy, or to active life.

I shall have much pleasure in presenting my wife to you and to Mrs. Purkis, on my return.

We have had a delightful tour through the Highlands. We are at the extreme point of our journey. The pleasures of a refined society — that of Lord and Lady Stafford's family — have induced us to make a long pause here. We think we shall be in London the beginning of December.

I have spent some days such as we passed together in Wales. We have had all the varieties of river, mountain, and wood scenery. The Lakes of

Scotland are infinitely finer than those of Wales; but the glens of the Principality may fairly stand in competition with those of the Highlands.

I hope I shall find you and your family in good health, and that you will have spent a very pleasant summer. I am, my dear Purkis,

Very sincerely and affectionately yours,

H. DAVY.

TO JOHN GEORGE CHILDREN, ESQ.

MY DEAR FRIEND, Dunkeld, Sept. 27, 1812.

I HAVE received your two kind letters. I hope your quiet life, and reasonable medical discipline, will entirely restore your health.

We are now on our return, and probably shall arrive in London before the middle of November: our time, however, is uncertain, as the Election may hasten, or keep us back for want of horses.

I can do nothing respecting the licence till my return; I will then see Mr. Wharton, or Mr. Vansittart. I have another subject of conversation in which they are interested, and I can easily introduce that of gunpowder.

I have been tolerably successful as a shot lately. I have not fished. My last adventure was at the Spey, near Gordon Castle, where I killed some noble salmon. At Blair Athol I shot some ptarmigans and a stag. I am now at Dunkeld, which I think the most beautiful habitable spot in the Highlands. The Tay, a noble river, rolls with a majestic stream through lofty woods seated upon cliffs and rounded hills; and in the background are the Mountains of Benyglor and the hills of Killycrankie.

My wife desires her kind remembrances. Pray offer mine to your father and daughter, and believe me to be always most affectionately yours,

H. DAVY.

TO THE SAME.

MY DEAR FRIEND, Edinburgh, October 14.

WE are on our return: I am well, but I am sorry to say that Lady D. is very much indisposed, and anxiety for her hastens my journey to town.

* * * * *

I have received a very interesting letter from Ampère. He says that a combination of chlorine and azote has been discovered at Paris, which is a fluid, and explodes by the heat of the hand; the discovery of which cost an eye and a finger to the author. He gives no details as to the mode of combining them. I have tried in my little apparatus with ammonia cooled very low, and chlorine, but without success.

There is little doing here. * * * * * * dresses and dances. Sir James Hall is writing on a sort of Deluge. Playfair is the true and amiable Philosopher. My brother is making experiments on animal matter.

I hope your gunpowder works are nearly finished. I shall be at the opening ball. As soon as I return I shall give my mind up to this matter. My wife desires her kind remembrances. Mine to your worthy father and Anna.

God bless you, my dear friend, and believe me

Ever affectionately yours,

H. DAVY.

On his return to town, after this tour, the following letter was addressed to his friend at Tonbridge:—

MY DEAR CHILDREN, October 24, 1812.

I HAVE just seen Pepys, and rejoice that he gives me so good an account of your health. My wife is much better, except that she has a swollen foot. I have never seen her in such good health and spirits. She is resolved to lead a home life of perfect quiet for six weeks, and I fear you will not be able to tempt her to quit her fire-side, though there is no visit she would make with greater pleasure: but lameness does not suit the country; and for one so enthusiastically fond of nature, it would be vexatious to be in the country, and not to be able to enjoy hills, and meads, and woods.

But I am ready to come to my business whenever you think I can be useful. I shall set to work to make gunpowder with as much ardour as Miles Peter—I hope with similar results.

I shall not be able to endure a very long separation from my wife, but for three or four days I am at your command.

I have been working yesterday and to-day on some new objects; and we are to have a meeting on Wednesday, at one o'clock, at the Institution, to try to make this compound of azote and chlorine, and to try some other experiments. Afterwards we (Angling Chemists) propose a dinner at Brunet's. If you can come to town on that day, I will promise to return with you.

God bless you, my dear Children, and believe me to be most affectionately yours,

H. DAVY.

CHAPTER IX.

Davy's "Elements of Chemical Philosophy" examined.—His Memoir on some combinations of Phosphorus and Sulphur, &c.—He discovers Hydro-phosphoric gas.—Important Illustrations of the Theory of Definite Proportionals.—Bodies precipitated from water are Hydrats.—His letter to Sir Joseph Banks on a new detonating compound.—He is injured in the eye by its explosion.—His second letter on the subject.—His paper on the Substances produced in different chemical processes on Fluor Spar.—His work on Agricultural Chemistry.

THE "Elements of Chemical Philosophy," a work to which he has alluded in several of the preceding letters, was published in June 1812. It is dedicated to Lady Davy, to whom he offers it "as a pledge that he shall continue to pursue Science with unabated ardour."

This work, although only a small part of the great labour he proposed to accomplish, must be considered as one of high importance to the cause of science. It has not perhaps announced any discoveries which had not been previously communicated to the Royal Society, but it has brought together his original results, and arranged them in one simple and digested plan—it has given coherence to disjointed facts, and has exhibited their mutual bearings upon each other, and their general relations to previously established truths.

Very shortly after the publication of this first part, it was asserted by a scientific critic that the work could never be completed upon the plan on which it had commenced, which was little less than a system of chemistry, in which all the facts were to be verified by the author: an undertaking far too gigantic for the most intrepid and laborious experimentalist to accomplish. There was too much truth in the remark:—the life of the Author has closed—the work remains unfinished.

Although it bears the title of "Elements," its plan and execution are rather adapted for the adept than the Tyro in science; it has, however, enabled the discoverer to expand several of his opinions with a freedom which is not consistent with the studied compression and elaborate brevity that necessarily characterise the style of a Philosophical Memoir,—and thus far it may have served the more humble labourer.

The first impression which this volume must produce, is that of admiration at the rapid and triumphant progress of Chemistry, during the period of a very few years; while a comparison of this work with others, even of very recent date, will show how much we are indebted for this progress to the unrivalled labours of Davy.

The first part of his projected system, which constitutes the volume under review, extends only to the general laws of chemical changes, and to the primary combinations of undecompounded bodies.

It is resolved into seven divisions, upon each of which I propose to offer some remarks.

THE FIRST DIVISION embraces the consideration of the three different forms of matter, *viz.* Solidity, Liquidity, and elastic Fluidity; and that of the active powers on which they depend, and by which they are changed, such as Gravitation, Cohesion, Calorific repulsion, or Heat, and Attractions chemical and electrical;—the laws of which he has expounded in a lucid and masterly manner; although it will be only necessary to quote the following passage, to show that the greatest philosopher may occasionally slide into error. "In solids, the attractive force predominates over the repulsive; in fluids, and in elastic fluids, they may be regarded as in different states of equilibrium; and in ethereal substances, the repulsive must be considered as predominating over and destroying the attractive force." A reviewer has very justly observed, that it is difficult to conceive how so much error and confusion could have been collected, by such an author, into so short a sentence. It is a solecism to say that two forces may exist in different states of equilibrium; besides, it is generally admitted that the repulsive force alone exists in elastic fluids, and that it is only compensated by external pressure, or gravitation.

In treating the subject of Heat, he maintains the same opinion, though in a manner somewhat more subdued, as that which he had formed at the very commencement of his scientific career,*—that it is

* See Page 44.

nothing else than motion, and that the laws of Heat are the same as the laws of Motion.

In taking a general view of the subject of Chemical Attraction, there is a remarkable clearness in his enunciation of its several propositions, and a great felicity in the selection of its illustrations. He combats the theory of Berthollet, respecting the influence of mass, with singular success, and confirms the general law, that all bodies combine chemically, in certain definite proportions to be expressed by numbers; so that, if one number be employed to denote the smallest quantity in which a body combines, all other quantities of the same body will be as multiples of this number; and the smallest proportions in which the undecompounded substances enter into union being known, the constitution of the compound they form may be learnt; and the element which unites chemically in the smallest quantity being expressed by unity, all the other elements may be represented by the relations of their quantities to unity. Unfortunately, however, there has existed amongst philosophers a want of agreement as to the *unit* to which the relative values of the other numbers shall be referred. Mr. Dalton selected Hydrogen as the unit; Davy followed his example, but doubled the weight of oxygen; while Wollaston, Thompson, and Berzelius, have proposed oxygen as the most convenient unit, since that element enters into the greatest number of combinations.

To Dalton is now universally conceded the glory of having established the laws of definite proportions; but in unfolding them, he has employed expressions which involve speculations as to their physical cause, and has thus given to that, which is nothing more than a copious collection of facts, the appearance of a refined theory. It may be perfectly true, as Mr. Dalton supposes, that all bodies are composed of ultimate atoms; but in the present state of our knowledge, we can neither form any idea of the nature of such atoms, nor of the manner in which they may be grouped together. We are therefore indebted to Davy for having, by his early and powerful example, taught the chemist how to disentangle fact from hypothesis, and to investigate the doctrine of proportionals, without any reference to the *atomic* theory which has been proposed for its explanation.

THE SECOND DIVISION treats of Radiant or Ethereal Matter, and of its effects in producing vision, heat, and chemical changes. It contains some refined speculations respecting the possible conversion of terrestrial bodies into light and heat, and *vice versâ*.

THE THIRD DIVISION presents us with an account of "Empyreal undecompounded Substances," or those which support combustion; together with that of the compounds which they form with each other. Upon this occasion, Davy has completely rescued us from the trammels of the Anti-phlogistic theory, and has shown that, so far from the process of combustion depending upon the position or transfer of oxygen, it is a *general* result of the actions of *any* substances possessed of strong chemical attractions, or different electrical relations, and that it takes place in all cases in which an intense and violent motion can be conceived to be communicated to the corpuscules of bodies, without any regard to the peculiar nature of the substances engaged. The announcement of the general law is followed by a history of the only two undecompounded bodies included under this arrangement, viz. *Oxygen*, and *Chlorine*.* In naming a class of bodies by their relations to combustion, he distinctly states that he merely intends to signify that the production of heat and light is more characteristic of their actions, than of those of any other substances; and that they are, at the same time, opposed to all other undecompounded substances by their electrical relations, being always in Voltaic combinations attracted to, or elicited from the positive surface; whereas all other known undecompounded substances are separated at the negative surface.

THE FOURTH DIVISION comprises the history of Undecompounded Inflammables, or Acidiferous Substances, not Metallic, and that of their binary combinations with oxygen and chlorine, or with each other.

The bodies considered under this division, are the following:—Hydrogen, Azote, Sulphur, Phosphorus, and Boracium, or Boron. Under the history of Sulphur, he gives us the true theory of the process by which sulphuric acid is produced by the combustion of that body in mixture with nitre, and which had never before been explained in any chemical work.

* *Iodine*, *Fluorine*, &c. had not been discovered at this period.

THE FIFTH DIVISION contains the Metals; their primary combinations with other undecompounded bodies, and with each other.

In the order of classification adopted on this occasion, the newly discovered inflammable metals, producing by combustion alkalies, alkaline earths, and earths, commence the series; next come those which produce oxides; and lastly, those which produce acids. Thus are we presented with a chain of gradations of resemblance which may be traced throughout the whole series of metallic bodies.

THE SIXTH DIVISION comprehends certain bodies (the *Fluoric Principle*, and the *Ammoniacal Amalgam)* which present some extraordinary and anomalous results. It is worthy of remark, that, at the period at which this work was written, Davy considered the peculiar acid developed from fluor spar, by the action of sulphuric acid, as a compound of an acid unknown in a separate state, and water; whence he proposed to call it *Hydro-fluoric* acid,—a term extremely objectionable from its ambiguity, since it would indicate either hydrogen or water as one of its constituents. At the conclusion, however, of this chapter, in consequence of having observed certain phenomena displayed by this gas, when in combination with silica and boracic acid, he for a moment seems to have caught the truth, but it as quickly eluded his grasp, and he dismisses the conjecture which it was his good fortune some years afterwards to verify, *viz.* that the fluoric acid is a compound of an unknown principle, analogous to chlorine, with hydrogen and water, and that *fluor spar* is a compound of the same principle with calcium, or the base of lime.

THE SEVENTH DIVISION offers to the chemical enquirer various speculations, as to the probable nature of certain bodies hitherto undecompounded. He observes, that "we know nothing of the true elements belonging to nature; but as far as we can reason from the relations of the properties of matter, that hydrogen is the substance which approaches nearest to what the elements may be supposed to be. It has energetic powers of combination, its parts are highly repulsive of each other, and attractive of the particles of other matter; it enters into combination in a quantity very much smaller than any other substance, and in this respect it is approached by no known body. After hydrogen, oxygen perhaps partakes most of the elementary character: it has a greater energy of attraction, and, with the exception just stated, enters into combination in the smallest proportion."

In conclusion, he hints at the possibility of the same ponderable matter in different electrical states, or in different arrangements, constituting substances chemically different, and he thinks that there are parallel cases in the different states in which bodies are found connected with their different relations to temperature: thus, steam, ice, and water, are the same ponderable matter; and certain quantities of steam and ice mixed together produce ice-cold water.

"That the forms of natural bodies may depend upon different arrangements of the same particles of matter, has been a favourite hypothesis, advanced in the earliest era of physical research, and often supported by the reasonings of the ablest philosophers. This sublime chemical speculation, sanctioned by the authority of Hooke, Newton, and Boscovich, must not be confounded with the ideas advanced by the alchemists, concerning the convertibility of the elements into each other. The possible transmutation of metals has generally been reasoned upon, not as a philosophical research, but as an empirical process. Those who have asserted the actual production of the precious metals, or their decomposition, or who have defended the chimera of the philosopher's stone, have been either impostors, or men deluded by impostors. In this age of rational enquiry, it will be useless to decry the practices of the adepts, or to caution the public against confounding the hypothetical views respecting the elements founded upon distinct analogies, with the dreams of alchemical visionaries, most of whom, as an author of the last century justly observed, professed an art without principles, the beginning of which was deceit, and the end poverty"

On the 18th of June 1812, Davy presented to the Royal Society a paper entitled "On some Combinations of Phosphorus and Sulphur; and on some other subjects of Chemical Inquiry."

By the researches detailed in this Memoir, he accomplished three important objects: he established the existence of some new compounds—furnished additional evidence in support of the doctrine of definite proportions—and ascertained that most of the substances obtained from aqueous solutions by precipitation, are compounds of water, or *Hydrats*. In the first place, he recognised the formation of two distinct compounds of phosphorus and chlorine: one, solid, white, and crystalline in its appearance;

the other, fluid, limpid as water, and volatile. The latter body he found to contain just double as much chlorine as the former.

On experimenting upon this latter body with water, he obtained a crystallized substance which he proposed to call *Hydro-phosphorous acid,* since it consists of pure phosphorous acid and water. By decomposition in close vessels, it is resolved into phosphoric acid, and a peculiar gas, consisting of one proportional of phosphorus and four of hydrogen, and for which he proposed the term *Hydro-phosphorous* gas. The reader, no doubt, will be immediately struck with the impropriety of a nomenclature in which the prefix *Hydro* is made to express water in the former, and hydrogen in the latter mstance.

In examining the results of the mutual decomposition of water and the phosphoric compounds of chlorine, Davy remarks, that it is scarcely possible to imagine more perfect demonstrations of the laws of definite combination: no products are formed except the new combinations, (phosphor*ic* acid from the solid, phosphor*ous* acid, from the liquid compound, and in both muriatic acid;) neither oxygen, hydrogen, chlorine, nor phosphorus, is disengaged; and therefore the ratio in which any two of them combine being known, the ratio in which the rest combine, in these cases, may be determined by calculation.

Lastly, he ascertained that most of the substances obtained by precipitation from aqueous solutions are compounds of water: thus zircona, magnesia, and silica, when precipitated and dried at 212°, still contain definite proportions of water; and many of the substances which had been considered as metallic oxides, he found, when obtained from solutions, to agree in this respect; and that their colours and other properties are materially influenced by this combined water.

On the 5th of November 1812, was read before the Royal Society a letter addressed by Davy to Sir Joseph Banks, on the subject of the detonating compound already alluded to in his communications to Mr. Children. He expresses his anxiety to have the circumstances made public as speedily as possible, since experiments upon the substance may be connected with very dangerous results.

He had some time before received information from Paris of a combination having been effected between chlorine and azote, and that it was distinguished by detonating properties; but he was wholly ignorant of the mode by which it had been prepared, and he could not obtain any information upon this point from any of the French journals.

So curious and important a result could not fail to interest him, as he had himself been long engaged in experiments on the action of azote and chlorine, without gaining any decided proofs of their power of combining with each other. It was evident from the notice, that this new body could not be formed in any operations in which heat is concerned; he therefore attempted to combine the elements by presenting them to each other artificially cooled, the azote being in a nascent state. For this purpose he introduced chlorine into a solution of ammonia; a violent action ensued, and minute films of a yellow colour were observed on the surface of the liquor, but they immediately resolved themselves into gas. As he was about to repeat the experiment with some other ammoniacal compounds, Mr. Children reminded him of the circumstance which he had previously communicated to him in a letter, that Mr. James Burton, junr, on exposing chlorine to a solution of nitrate of ammonia, had observed the formation of a yellow oil, but which he had not been able to collect. Davy availed himself of the hint, and obtained the substance in question: on examining its properties by the application of heat, the tube in which it was contained was shivered to atoms by its explosion, and he received a severe wound in the transparent cornea, which was followed by inflammation, and disabled him from pursuing his enquiry.

In the following July, however, he communicated in a second letter to Sir Joseph Banks, the continuation of this enquiry, and furnished a full and satisfactory history of the body in question. Having procured it in sufficient quantity, he attempted to effect its analysis by the action of mercury, but a violent detonation occurred, and he was again wounded in the head and hands; fortunately, however, the injury was slight, in consequence of his having taken the precaution to defend his face by a plate of glass attached to a proper cap.

In a subsequent experiment, by using smaller quantities, and recently distilled mercury, he succeeded in obtaining results without any violence of action: the mercury united with the chlorine, and the azote was disengaged; from which he was enabled to conclude that it was composed of four volumes of chlorine and one volume of azote. For this new body Davy suggested the name of *Azotane;* but I have already observed, that his nomenclature of the compounds of chlorine has never been adopted; the detonating substance is now very properly denominated *Chloride of Nitrogen.*

Shortly after the publication of this paper, M. Berzelius, in a letter to Professor Gilbert, asserted that "*Azotane*" is nothing more than *dry* nitro-muriatic acid, since it dissolves slowly in water, and forms a weak *aqua regia.* "These few observations," says he, "show clearly that Davy's analysis of this substance is inaccurate, and that he corrected his results in consequence of theoretical views."

This was an imputation upon the philosophical character of Davy, which excited in him no small degree of indignation. In reply he says, "It is difficult to discover what meaning M. Berzelius attaches to the term *dry* nitro-muriatic acid; and it is wholly unnecessary to refute so unfounded and vague an assertion."

On July 8, 1813, a paper was read by Davy before the Royal Society, entitled "Some Experiments and Observations on the Substances produced in different chemical processes on Fluor Spar."

The views which he formerly entertained with respect to the fluoric acid have been already noticed:* in the present paper he renounces his previous opinions, and establishes, by experiments of the most satisfactory character, that the base of fluoric acid is a highly energetic body not hitherto obtained in an insulated form, and the properties peculiar to which

* See page 364.

are as yet unknown. It appears, however, to belong to the class of negative electrics, and, like oxygen and chlorine, to have a powerful affinity for hydrogen and metallic substances. With hydrogen, it constitutes the peculiar and very powerful acid long known by the name of *fluoric acid*,—with boron, the *fluoboric*, and with silicium, the *silicated-fluoric*, acids. Although this theory had originally suggested itself to the mind of Davy, yet the chemical world is unquestionably indebted to M. Ampère for establishing it; and the English chemist has very justly acknowledged the obligation. "During the period that I was engaged in these investigations," says he, "I received two letters from M. Ampère, of Paris, containing many ingenious and original arguments in favour of the analogy between the muriatic and fluoric compounds. M. Ampère communicated his views to me in the most liberal manner: they were formed in consequence of my ideas on chlorine, and supported by reasonings drawn from the experiments of MM. Gay Lussac and Thénard."

It has been stated that Davy gave his last public lecture on the 9th of April 1812; he however afterwards delivered an occasional lecture to the Managers, on his own discoveries, and did not formally resign his professorship until the next year.

The following record has been extracted from the Journal of the Institution.

"Minutes of the Proceedings of a general Monthly Meeting of the Members of the Royal Institution, held on Monday, April 5, 1813.

"Earl of Winchelsea, President, in the Chair.

"This being the meeting appointed by Article 2.

chap. xix. of the bye-laws, for putting in nomination from the chair the professors for the year ensuing, Sir Humphry Davy rose, and begged leave to resign his situation of Professor of Chemistry; but he by no means wished to give up his connection with the Royal Institution, as he should ever be happy to communicate his researches, in the first instance, to the Institution, in the manner he did in the presence of the members last Wednesday, and to do all in his power to promote the interest and success of this Institution.

"Sir H. Davy having retired, Earl Spencer moved, That the thanks of this Meeting be returned to Sir H. Davy, for the inestimable services rendered by him to the Royal Institution. This motion was seconded by the Earl of Darnley, and on being put, was carried unanimously.

"Earl Spencer further moved, That in order more strongly to mark the high sense entertained by this Meeting of the merits of Sir H. Davy, he be elected Honorary Professor of Chemistry; which, on being seconded by the Earl of Darnley, met with unanimous approbation.

"The Chairman having declared the Professorship of Chemistry vacant, put in nomination William Thomas Brande, Esq. F.R.S. as a candidate for that office, with a salary of 200*l.* per annum.

"On Monday, June 7, 1813, William Thomas Brande, Esq. was unanimously elected."

In March 1813, Davy published his "Elements of Agricultural Chemistry," being the substance of a course of lectures which he had, for ten successive seasons, delivered before the members of the Board of Agriculture, to whom the work is inscribed, as a mark of the author's respect.

This work, which may be considered as the only system of philosophical agriculture ever published in this country, has not only contributed to the advancement of science, but to that for which he has an equal claim upon our gratitude,—the diffusion of a taste amongst the higher classes for its cultivation; for it has been wisely remarked, that not he alone is to be esteemed a benefactor to mankind who makes an useful discovery, but he, also, who can point out an innocent pleasure.

It has been already stated, that Davy became early impressed with the importance of the subject: —that in future life its investigation should have been to him so fertile a source of pleasure, may be readily imagined, when it is remembered with what passionate delight he contemplated the ever varying forms of creation. "I am," said he, "a lover of Nature, with an ungratified imagination, and I shall continue to search for untasted charms—for hidden beauties." In unfolding, then, the secrets of vegetable life, he did but remove the veil from his mistress. From the same poetical feeling sprang his love of angling: it was a pursuit which carried him into the wild and beautiful scenery of Nature, amongst the mountain lakes, and the clear and lovely streams that gush from elevated hills, or make their way through the cavities of calcareous strata.* In the early spring, it led him forth upon the fresh turf in the vernal sunshine, to scent the odour of the bank perfumed by the violet, and

* See his Salmonia, Edit. 2. p. 9.

enamelled with the primrose, while his heart participated in the renovated gladness of Nature.

I had hoped that, amidst the voluminous correspondence of my late friend Mr. Arthur Young, some important letters might have been found from Davy on agricultural subjects; but the communications which took place between them were generally in conversation, and I have therefore only been able to procure two letters, which I shall here insert: the first will show that, during his tours, his attention was alive to the practices of husbandry; and the second will prove that he had once seriously contemplated the labour of writing the agricultural history of his native county.

TO ARTHUR YOUNG, ESQ.

DEAR SIR, Killarney, June 1806.

YOU have been of great and durable service to Ireland. I have met with a number of persons who have been enlightened by your labours, and who now follow an enlightened system of Agriculture. One very intelligent gentleman you will recollect,—Mr. Bolton of Waterford: he is zealously pursuing improvements, and is instructing his neighbours by precept and example. I am, &c. H. DAVY.

The above letter contains also some observations on a chemical mixture, but which is unintelligible from our being ignorant of the conversation to which it refers.

TO THE SAME.

DEAR SIR, April, 1807.

I CALLED this morning with the hope of seeing you, and of gaining some explanation on the subject of your note. I shall not be able to leave London until the middle of July, and I must return early in October.

I do not think there would be sufficient time between these periods for accomplishing the objects you mention; nor do I think myself qualified to write upon the agriculture of a county. I wished likewise to devote the leisure of this summer to the preparation of my lectures on the Chemistry of Agriculture for publication. I have a great deal of information concerning the mineralogy and geology of Cornwall, but none concerning the farming.

If the business admits of being postponed, I might perhaps be able to accomplish it next summer; that is, by devoting a part of this summer, and the whole of my next: but I would rather confine myself to my own province, the mineralogy and geology of the county, and leave the agriculture to abler hands.

Be pleased to receive my thanks, and to communicate them to the President for the honour of the proposal. I remain, &c.

H. Davy.

The majority of my readers will probably concur in the wisdom of this decision: they will consider that to have doomed Davy to a drudgery of this nature, would have been wasting talents upon an object which might be accomplished by smaller means. From my acquaintance, however, with Cornwall, I am induced to form a different opinion. Davy never approached even those subjects which had already received from others the most thorough investigation, without extracting from them new and important truths. What, then, might not have been expected from his genius, when applied to a department upon which the light of science had scarcely dawned?

It is only in a primitive country like Cornwall, that the natural relations between the varieties of soil and the subjacent rocks can be studied with success: as we advance to alluvial districts, such relations become gradually less distinct and apparent, and are ultimately lost in the confused complication of the soil itself, and in that general obscurity which envelopes every object in the ulterior stages of decomposition. We can, therefore, only hope to succeed in such an investigation by a patient and laborious examination of a primitive country, after which we may be enabled to extend our enquiries with greater advantage through those regions which are more completely covered with soil, and obscured by luxuriant vegetation; as the eye, acquainted with the human figure, on gazing upon a beautiful statue, traces the outline of the limbs, and the swelling contour of its form, through the flowing draperies which invest it. The importance of the subject, as well as the general interest it has excited, induce me to offer an analysis of his "Elements of Agricultural Chemistry."

The work is divided into eight lectures; and in his introductory chapter, after adverting to the difficulties which the enquiry presents to the lecturer, he offers a general view of the objects of the course, and of the order in which he proposes to discuss them.

"Agricultural Chemistry has not yet received a regular and systematic form. It has been pursued by competent experimenters for a short time only; the doctrines have not as yet been collected into an elementary treatise; and on an occasion when I am obliged to trust so much to my own arrangements, and to my own limited information, I cannot but feel diffident as to the interest that may be excited, and doubtful of the success of the undertaking. I know, however, that your candour will induce you not to expect any thing like a finished work upon a science as yet in its infancy; and I am sure you will receive with indulgence the first attempt made to illustrate it, in a distinct course of lectures.

"Agricultural Chemistry has for its objects all those changes in the arrangements of matter connected with the growth and nourishment of plants; the comparative values of their produce as food; the constitution of soils; and the manner in which lands are enriched by manure, or rendered fertile by the different processes of cultivation." That such objects are intimately connected with the doctrines of chemistry, he proceeds to show by several appropriate and striking illustrations.

"If land be unproductive, and a system of ameliorating it is to be attempted, the sure method of obtaining the object is, by determining the cause of its sterility, which must necessarily depend upon some defect in the constitution of the soil, which may be easily discovered by chemical analysis. Are any of the salts of iron present? they may be decomposed by lime. Is there an excess of siliceous sand? the system of improvement must depend on the application of clay and calcareous matter. Is there a defect of calcareous matter? the remedy is obvious. Is an excess of vegetable matter indicated? it may be removed by liming, paring, and burning. Is there a deficiency of vegetable matter? it is to be supplied by manure."

"In the selection also of the remedy, after the discovery of the evil, chemical knowledge is of the highest importance. Limestone varies in its composition, and by its indiscriminate application we may aggravate the sterility we seek to obviate. Peat earth is an excellent manure, but it may contain such an excess of iron as to be absolutely poisonous to plants. How are such difficulties to be met but by the resources of chemistry? It is also evident that the scientific agriculturist should possess a general knowledge of the nature and composition of material bodies, and the laws of their changes; for the surface of the earth, the atmosphere, and the water deposited from it, must, either together or separately, afford all the principles concerned in vegetation; and it is only by examining the chemical nature of these principles, that we are capable of discovering what is the food of plants, and the manner in which this food is supplied and prepared for their nourishment."

Davy likewise advocates the necessity of studying "the phenomena of vegetation, as an important branch of the science of organized nature; for, although exalted above inorganic matter, vegetables are yet in a great measure dependent for their existence upon its laws. They receive their nourishment from the external elements; they assimilate it by means of peculiar organs; and it is by examining their physical and chemical constitution, and the substances and powers which act upon them, and the modifications which they undergo, that the scientific principles of Agricultural Chemistry are obtained."

With respect, however, to the practical utility of this latter branch, different opinions have been entertained. I confess, I am inclined to agree with an able reviewer* when he says, "It is the proper business of the chemist to examine and ascertain the nature and properties of dead and inorganized matter, and the various combinations which, according to chemical laws, it is capable of forming. The chemical composition of organized bodies, and of the products which they form, fall likewise under his cognizance; but when he proceeds to consider the physical constitution of these bodies, and the manner in which they act in forming their products, he no longer works with the instruments of the laboratory, or conducts processes which can be properly imitated there."

In concluding his introductory observations, he remarks upon the prejudice which persons, who argue in favour of practice and experience, very commonly entertain against all attempts to improve agriculture by philosophical enquiries and chemical methods. "That much vague speculation may be found in the works of those who have lightly taken up agricultural chemistry, it is impossible to deny. It is not uncommon to find a number of changes rung upon a string of technical terms, such as oxy-

* Edinburgh Review, vol. 22, page 253.

gen, hydrogen, carbon, and azote, as if the science depended upon words, rather than upon things. But this is, in fact, an argument for the necessity of the establishment of just principles of chemistry on the subject.—If a person journeying in the night wishes to avoid being led astray by the ignis fatuus, the most secure method is to carry a lamp in his own hand."

"There is no idea more unfounded than that a great devotion of time, and a minute knowledge of general chemistry, are necessary for pursuing experiments on the nature of soils, or the properties of manures. The expense connected with chemical enquiries is extremely trifling: a small closet is sufficient for containing all the materials required."

In the SECOND LECTURE, he enters upon the consideration of the general powers of matter, such as gravitation, cohesion, chemical attraction, heat, light, and electricity; and then proceeds to examine the elements of matter, and the laws of their combinations and arrangements.

To an audience constituted of persons who were not familiar with the elementary principles of the science, it might have been very necessary for the lecturer to enter upon such preliminary details; but there cannot be any good reason for his having published them in his system. As they are to be found in every work on chemistry, it will not be necessary to bestow upon them any further notice.

In the THIRD LECTURE, he enters into a description of the organization and living system of plants; in which he connects together into a general view, the observations of the most enlightened philosophers who have studied the physiology of vegetation —those of Grew, Malpighi, Sennebier, Hales, Decandolle, Saussure, Bonnet, Darwin, Smith, and above all, of Mr. Knight, whose enquiries upon these subjects are not only the latest, but by far the most satisfactory and conclusive.

As there is little in these descriptions that may not be found in the original authors, I shall not unnecessarily trespass upon the time of the reader by relating them. In the latter part of this lecture, he describes the properties and ultimate composition of the proximate principles of which vegetable matter consists, and into which it may be resolved by different processes of art; such are gum, starch, sugar, albumen, gluten, extract, tannin, resin, oils, &c. &c. But since the publication of this work, vegetable analysis has advanced to a degree of refinement which could scarcely have been anticipated in so short a period, and consequently many of his statements appear deficient; but his general directions for conducting an analysis of any vegetable substance, with a degree of accuracy sufficient for the views of the agriculturist, remain unimpeached.

The most valuable, and more strictly original part of this lecture, is his statement of the quantity of soluble or nutritive matters contained in varieties of the different substances that are used as articles of food, either for man or cattle, and which he has displayed in a tabular form.

The analyses were his own, and were conducted with a view to a knowledge of the general nature and quantity of the products, rather than to that of their intimate chemical composition. He proceeded upon the assumption, that the excellence of the different articles, as food, will be in a great measure proportional to the quantities of soluble matter they afford; although he admits that these quantities cannot be regarded as *absolutely* denoting their value. Albuminous or glutinous matters have the characters of animal substances; sugar is more, and extractive matter less nourishing than any other principles composed of carbon, hydrogen, and oxygen. Certain combinations likewise of these substances may be more nutritive than others. There are some principles also, which, although soluble in the vessels of the chemist, pass through the alimentary canal of animals without change; such is *tannin:* on the other hand, there are bodies which, although sparingly soluble in water, are readily acted upon by the gastric juice; *gluten* is a principle of this description.

Shortly after Dr. Wollaston published his scale of chemical equivalents, it occurred to me that by applying the sliding rule to a series of nutritive substances, arranged according to the analyses of Davy, some curious and important problems * might be solved; or at least, that the accuracy of the conclusions might be thus conveniently submitted to the test of practice. I accordingly superintended the construction of such an instrument, and submitted it to Davy, who expressed his approbation of the principle, but doubted how far the accuracy of his analyses would justify the experiment.

* For example:—What weight of wheat is equivalent to a given weight of oats, barley, rye, &c.? Suppose three hundred pounds of potatoes feed twenty head of cattle for any given time, how many will the same weight of oats feed?

To such a scheme, however, I soon found that there existed a much more serious objection. The operation of the insoluble matter had been wholly neglected; and whatever views the chemist may entertain, the experience of the physiologist has established, beyond doubt, the influence of such matter in the process of digestion. The capacity of the alimentary organs of graminivorous animals sufficiently proves that they were designed for the reception of a *large bulk* of food, and not for provender in which the nutritive matter is concentrated; and since the gramineous and leguminous vegetables do not present this matter in a separate state, and the animal is not furnished with an apparatus by which he can remove it, the obvious inference is, that he was designed to feed indiscriminately upon the whole; and that, unless bulk be taken into the account, no fair inference can be deduced as to the nutritive value of different vegetables.

Notwithstanding the difficulties which prevent our arriving at any thing like an accurate conclusion upon so complicated a subject, the results may be received as affording some general views with regard to the comparative value of different nutritive vegetables. It would thus appear that at least a fourth part of the weight of the potatoe consists of nutritive matter, which is principally starch;—that wheat consists of as much as ninety-five, barley of ninety-two, oats of seventy-five, rye of eighty, and peas and beans of about fifty-seven per cent. of nutritive matter.

The FOURTH LECTURE comprises subjects of the utmost importance, and must be considered as constituting by far the most original and valuable division of the work. It treats of soils,—their constituent parts, their chemical analysis, their uses, their improvement, and of the rocks and strata found beneath their surface.

In the execution of this part of his labours, he has not only improved on the processes of Fordyce and Kirwan, but he has enriched the subject with much interesting and novel research.

"Soils, although extremely diversified in appearance and quality, consist of comparatively few elements, which are in various states of chemical combination, or of mechanical mixture.

"These substances are silica, lime, alumina, magnesia, the oxides of iron, and of manganese; animal and vegetable matters in a state of decomposition; together with certain saline bodies, such as common salt, sulphate of magnesia, sometimes sulphate of iron, nitrates of lime and magnesia, sulphate of potash, and the carbonates of potash and soda.

"The silica in soils is usually combined with alumina and oxide of iron; or with alumina, lime, magnesia, and oxide of iron, forming gravel and sand of different degrees of fineness. The carbonate of lime is usually in an impalpable form; but sometimes in the state of calcareous sand. The magnesia, if not combined in the gravel and sand of the soil, is in a fine powder united to carbonic acid. The impalpable part of the soil, which is commonly called clay or loam, consists of silica, alumina, lime, and magnesia; and is, in fact, usually of the same composition as the hard sand, but more finely divided. The vegetable, or animal matters (and the first is by far the most common in soils,) exist in different states of decomposition. They are sometimes fibrous, sometimes entirely broken down and mixed with the soil.

"To form a just idea of soils, it is necessary to conceive different rocks decomposed, or ground into parts and powder of different degrees of fineness; some of their soluble parts dissolved by water, and that water adhering to the mass, and the whole mixed with larger or smaller quantities of the remains of vegetables and animals, in different stages of decay."

Soils, then, would appear to have been originally produced from the disintegration of rocks and strata; and hence there must be at least as many varieties of them, as there are species of rocks exposed at the surface of the earth; and they may be distinguished by names derived from the rocks from which they were formed. Thus, if a fine red earth be found immediately above decomposing basalt, it may be denominated *basaltic* soil. If fragments of quartz and mica be found abundant, it may be denominated *granitic* soil; and the same principles may be extended to other analogous cases.

A general knowledge then of geology becomes essential to the scientific agriculturist, not only to enable him to form a correct judgment with respect to the connection between the varieties of soil and the subjacent rocks, but to direct him to the different mineral substances which may be associated together in their vicinity, and which may contain principles capable of extending their fertility, or of correcting the circumstances upon which their poverty or barrenness may depend.

With this conviction, Davy proceeds to offer a general view of the nature and position of rocks and strata in nature; but which, I confess, appears to me to be wholly useless to those who have any acquaintance with the subject, and far too meagre to convey any instruction to those who have not made this branch of science an object of study.

Upon this view, however, he has grounded a number of valuable remarks; although his observations appear to have been too limited to enable him to do justice to a subject of such extent and importance. Had he fulfilled his intention of making a survey of the county of Cornwall, the science must have been greatly advanced by his labours, for there is no district in Great Britain so rich in fact, and so capable of elucidating the history of soil, and the advantages of cultivation, when conducted on the principles of chemical philosophy. The soils superincumbent upon the different rocks are distinct and characteristic; and even in the same species varieties may be observed, in consequence of geological peculiarities. I have, for instance, found that the fertility of a granitic soil is increased by the abundance of felspar in the parent rock;— that of a slaty soil by the degree of inclination or dip of the strata: but the most extraordinary circumstance perhaps connected with this subject, is the very remarkable fertility of the land which lies over the junction of these rocks,—so obvious indeed is it, that the eye alone is sufficient to trace it.

We are indebted to the author, in this lecture,

for some very ingenious and important remarks on the relations of different soils to heat and moisture, and for a series of experiments by which his views are supported.

Some soils, he observes, are more easily heated and more easily cooled than others: for example, those that consist principally of a stiff white clay are heated with difficulty; and being usually very moist, they retain their heat only for a short time. *Chalks* also are difficultly heated; but being dryer, they retain their heat longer, less being consumed in the process of evaporation.

A black soil, and those that contain much carbonaceous or ferruginous matter, acquire a higher temperature by exposure to the sun, than pale-coloured soils.

When soils are perfectly dry, those that most readily become heated, most rapidly cool; but the darkest-coloured dry soil, abounding in animal and vegetable matters, cools more slowly than a wet pale soil, composed entirely of earthy matter.

These results Davy gained by experiments made on different kinds of soils, exposed for a given time to the sun, and in the shade; the degrees of heating and cooling having been accurately ascertained by the thermometer.

Nothing can be more evident, than that the genial heat of the soil, particularly in spring, must be of the highest importance to the rising plant. And when the leaves are fully developed, the ground is shaded, and any injurious influence, which in the summer might be expected from too great a heat, entirely prevented; so that the temperature of the surface, when bare and exposed to the rays of the sun, affords at least one indication of the degree of its fertility; and the thermometer may therefore be sometimes a useful instrument to the purchaser or improver of lands.

Water is said to exist in soils, either in a state of chemical combination, or of cohesive attraction. It is in the latter state only that it can be absorbed by the roots of plants, unless in the case of the decomposition of animal and vegetable substances. The more divided the parts of the soil are, the greater is its attractive power for water; and the addition of vegetable and animal matters still farther increases this power.

The quality of soils to absorb water from air, is much connected with fertility. Davy informs us that he has compared this absorbent power in numerous instances, and that he always found it greatest in the most productive lands: he states, however, the important fact, that those soils, such for instance as stiff clays, which take up the greatest quantity of water, when it is poured upon them in a fluid form, are not such as absorb most moisture from the atmosphere in dry weather. They cake, and present only a small surface to the air, and the vegetation on them is generally burnt up almost as readily as on sands.

There is probably no district in which the importance of moisture in relation to fertility is more apparent than in Cornwall; and there is a provincial saying, that the land will bear a shower every weekday, and two upon a Sunday: indeed, of such importance is moisture, that it is by no means an uncommon practice to encourage the growth of weeds, in order to diminish the evaporation; a necessity which arises from the excess of siliceous matter in the soil.

To those who are disposed to prosecute this enquiry, I should recommend a perusal of Mr. Leslie's treatise on the "Relations of Air to Heat and Moisture."

I must not quit the consideration of this lecture, without adverting to the directions with which its author has furnished the philosophical farmer for analysing the different varieties of soil; and which are so clear, so perfect, and above all so simple, that they are now introduced into all elementary works on chemistry, as the only guide to such researches. His method for ascertaining the quantity of carbonate of lime in any specimen, consists in determining the loss of weight which takes place on its admixture with muriatic acid; for since carbonate of lime, in all its states, contains a determinate proportion of carbonic acid, it is evident that, by estimating the quantity of elastic matter given out, the proportion of carbonate of lime will be known. For conducting this experiment, he contrived a very simple and ingenious piece of pneumatic apparatus, in which the bulk of the carbonic acid is at once measured by the quantity of water it displaces.

In his FIFTH LECTURE he enters upon the nature of the atmosphere, and its influence on vegetables: he also examines the process of the germination of seeds, and the functions of plants in their different stages of growth; and concludes with a general view of the progress of vegetation.

I shall merely mention a few of the more interesting points in this enquiry.

In illustrating the importance of water to the vegetable creation, he observes that the atmosphere always contains water in its elastic and invisible form, the quantity of which will vary with the temperature. In proportion as the weather is hotter, the quantity is greater; and it is its condensation by diminution of temperature, which gives rise to the phenomena of dew and mist. The leaves of living plants appear to act upon this vapour, and to absorb it. Some vegetables increase in weight from this cause, when suspended in the atmosphere, and unconnected with the soil; such are the house-leek, and different species of the aloe. In very intense heats, and when the soil is dry, the life of plants seems to be preserved by the absorbent powers of their leaves; and it is a beautiful circumstance in the economy of Nature, that aqueous vapour is most abundant in the atmosphere when it is most needed for the purposes of life; and that when other sources of its supply are cut off, this is most copious.*

* The history of his native county would have furnished him with a parallel instance of the intelligence and design which Nature displays in connecting the wants and necessities of the different parts of creation, with the power and means of supplying them. In a primitive country like Cornwall, the siliceous soil necessarily requires much moisture, and we may perceive that the cause which occasions, at the same time supplies this want; for the rocks elevated above the surface, solicit a tribute from every passing cloud; while in alluvial and flat districts, where the soil is rich, deep, and retentive of moisture, the clouds float undisturbed over the plains, and the country frequently enjoys that uninterrupted series of dry weather which is so necessary to its fertility.

If water in its elastic and fluid states be essentially necessary to the economy of vegetation, so even in its solid form, it is not without its uses. Snow and ice are bad conductors of heat; and at a period when the severity of the winter threatens the extinction of vegetable life, Nature kindly throws her snowy mantle over the surface; while in early spring the solution of the snow becomes the first nourishment of the plant; at the same time, the expansion of water in the act of congelation, and the subsequent contraction of its bulk during a thaw, tend to pulverise the soil, to separate its parts from each other, and, by making it more permeable to the influence of the air, to prepare it for the offices it is destined to perform.

He next proceeds to consider the action of the atmosphere on plants, and to connect it with a general view of the progress of vegetation. He commences with examining its relations to germination.

"If a healthy seed be moistened and exposed to air at a temperature not below 45°, it soon germinates; it shoots forth a *plume* which rises upwards, and a *radicle* which descends.

"If the air be confined, it is found that, in the process of germination, the oxygen, or a part of it, is absorbed. The azote remains unaltered; no carbonic acid is taken away from the air; on the contrary, some is added." Upon this point, critics have been disposed to break a lance with Sir Humphry.

Linnæus observes, that the plants which chiefly grow upon the summit of mountains, are rarely found in any other situation, except in marshes, because the clouds arrested in their progress by such elevations, keep the air in a state of perpetual moisture.

The doctrine, let it be observed, is at variance with the numerous experiments made on this subject by Scheele, Cruickshank, and De Saussure; the results of which agree in proving, that if seeds be confined and made to germinate in a given portion of air, not a *part* only, but the *whole* of the oxygen is consumed; and that its place is supplied, not merely by *some*, but by an *equal bulk* of carbonic acid.

Objections have been also started to his theory of the chemical changes which the seed undergoes during the process of germination: but were I to enter upon these discussions, time and space would alike fail me, to say nothing of the patience of the reader, which would be exhausted long before we could arrive at any satisfactory conclusion. I shall for the same reasons pass over his observations upon the influence exerted upon growing plants on the air: the subject is involved in much difficulty, which can be only removed by fresh experiments; nor, after all, is the great question, whether the purity of the atmosphere is maintained by vegetation, of any practical moment,—it is one which partakes more of curiosity than of use, and might therefore have been well dispensed with in a system of agriculture.

He agrees with many other philosophers in considering "the process of malting as merely one in which germination is artificially produced, and in which the starch is changed into sugar, which sugar is afterwards, by fermentation, converted into spirit.

"It is," he continues, "very evident from the chemical principles of germination, that the process should be carried on no farther than to produce the sprouting of the radicle, and should be checked as soon as this has made its distinct appearance. If it is pushed to such a degree as to occasion the perfect developement of the radicle and the plume, a considerable quantity of saccharine matter will have been consumed in producing their expansion, and there will be less spirit formed in fermentation, or produced in distillation.

"As this circumstance is of some importance, I made, in October 1806, an experiment relating to it. I ascertained by the action of alcohol, the relative proportions of saccharine matter in two equal quantities of the same barley; in one of which the germination had proceeded so far as to occasion protrusion of the radicle to nearly a quarter of an inch beyond the grain in most of the specimens, and in the other of which it had been checked before the radicle was a line in length; the quantity of sugar afforded by the last was to that in the first nearly as six to five."

The whole of this subject appears to be debateable ground between the physiologists and chemists: the one considering the change of starch into sugar as the result of the vital action of the seed; the other affirming that the growth of the germ is in no way necessary to the result, and is to be considered as a mere indication of the due degree of change being effected in the organic matter, or, in other words, that when the organized parts exhibit a certain degree of developement, then the inorganic matter is most completely changed. All growth beyond this is injurious, as leading to a consumption of the inorganic matter. All less than this is not otherwise disadvantageous, than as an indication that the inorganic matter is not duly changed. This change, it is farther affirmed, so far from depending upon vegetable life, can be wrought on the matter of the seed after it is even reduced to powder, or is separated in the form of starch. At all events, it must be admitted as a beautiful arrangement in nature, that the same agents which urge on the developement of the organized parts, should, at the same time, assist in preparing food for their support.

From this subject Davy is very naturally led to the consideration of the ravages inflicted upon the infant plant by insects; the saccharine matter in the cotyledons at the time of their change into seed-leaves, rendering them exceedingly liable to such attacks. He appears to have bestowed much attention on the turnip-fly, a colyopterous insect, which fixes itself upon the seed-leaves of the turnip at the time that they are beginning to perform their functions. He relates the several remedies which have been proposed for this evil; and from letters which have been put in my possession, addressed to Dr. Cartwright as early as the year 1804, he appears to have been engaged with that gentleman in experiments made by sprinkling the young plants with lime and urine.

After alluding to the parasitical plants of different species, which attach themselves to trees and shrubs, feed on their juices, destroy their health, and finally their life, for which, at present, there does not exist any remedy, he thus concludes his lecture:

"To enumerate all the animal destroyers, and tyrants of the vegetable kingdom, would be to give a catalogue of the greater number of the classes in Zoology. Every species of plant almost is the peculiar resting-place, or dominion, of some insect tribe; and from the locust, the caterpillar, and snail, to the minute aphis, a wonderful variety of the inferior insects are nourished, and live by their ravages upon the vegetable world.

"The Hessian fly, still more destructive to wheat than the one which ravages the turnip plant, has in some seasons threatened the United States with a famine. And the French government is at this time* issuing decrees with a view to occasion the destruction of the larvæ of the grasshopper.

"In general, wet weather is most favourable to the propagation of mildew, funguses, rust, and the small parasitical vegetables; dry weather, to the increase of the insect tribes. Nature, amidst all her changes, is continually directing her resources towards the production and multiplication of life; and in the wise and grand economy of the whole system, even the agents that appear injurious to the hopes, and destructive to the comforts of man, are in fact ultimately connected with a more exalted state of his powers and his condition. His industry is awakened, his activity kept alive, even by the defects of climates and season. By the accidents which interfere with his efforts, he is made to exert his talents, to look farther into futurity, and to consider the vegetable kingdom, not as a secure and unalterable inheritance spontaneously providing for his wants, but as a doubtful and insecure posses-

* January 1813.

sion, to be preserved only by labour, and extended and perfected by ingenuity."

His SIXTH LECTURE treats of manures of animal and vegetable origin, and of the general principles with respect to their uses and modes of application.

It is evident that plants, by their growth, must gradually exhaust the soil of its richer and more nutrient parts; and these can be alone restored by the application of manures. It is equally obvious, that if a soil be sterile from any defect in its constitution, such a defect can be only remedied by artificial additions. Hence the introduction of foreign matter into the earth, for the purpose of accelerating vegetation, and of increasing the produce of its crops, is a practice which has been pursued since the earliest period of agriculture. Unfortunately, however, the greatest ignorance has prevailed in all ages with regard to the best modes of rendering such a resource available; and the farmer, instead of enriching the soil, has too frequently given his treasures to the winds. "It is quite lamentable," says an intelligent writer,* "to survey a farm-yard in many parts of the kingdom; to see the abundance of vegetable matter that is trodden for months under-foot, over a surface of perhaps half an acre of land, exposed to all the rains that fall, by which its more soluble and richer parts are washed away, or perhaps carried down to poison the water of some stagnant pool, which the unfortunate cattle are afterwards compelled to drink. From the yard, the manure is often carted to the field, at the time when the land is rendered impenetrable by frost; or, if this opera-

* Edinburgh Review, vol. xxii. p. 270.

tion be delayed to a less unseasonable period, it is then frequently laid down in small heaps, or sometimes spread over the surface, exposed for many days to the sun, the winds, and the rain, as if with the direct design of dissipating those more volatile parts which it ought to be the farmer's first endeavour to preserve.

"Nothing can be so likely to remove ignorance so deplorable, and prejudices so inveterate, as the diffusion of real knowledge concerning the nature of manures, and their mode of action on soils, and on the plants which grow in them."

Davy, fully sensible of the practical importance of the subject, and impressed with the conviction that it was capable of being materially elucidated by the recent discoveries of chemistry, determined to put forth his strength, in order to bring this department of agriculture under the dominion of science; and upon this occasion our philosopher presents himself in the only character in which he ever ought to appear—in that of an original experimentalist.

His first step in the enquiry was to ascertain whether solid substances can pass from the soil through the minute pores in the fibres of the root. He tried an experiment by introducing a growing plant of peppermint into water which held in suspension a quantity of impalpably powdered charcoal: but after a fortnight, upon cutting through different parts of the roots, no carbonaceous matter could be discovered in them, nor were the smallest fibres even blackened,—though this must have happened, had the charcoal been absorbed in a solid form. If a substance so essential to plants as carbonaceous matter cannot be introduced except in a state of solution into their organs, he very justly concludes that other less essential bodies must be in the same case.

He also proved by experiment that solutions of sugar, mucilage, jelly, and other principles, unless considerably diluted, clogged up the vegetable organs with solid matter, and prevented the transpiration by the leaves: when, however, this precaution was taken, the plants grew most luxuriantly in such liquids.

He next proceeded to determine whether soluble vegetable substances passed in an unchanged state into the roots of plants, by comparing the products of the analysis of the roots of plants of mint which had grown, some in common water, some in a solution of sugar: the results favoured the opinion that they were so absorbed. It appeared, moreover, that substances even poisonous to vegetables did not offer an objection to this law. He introduced the roots of a primrose into a weak solution of oxide of iron in vinegar, and suffered them to remain in it till the leaves became yellow; the roots were then carefully washed in distilled water, bruised, and boiled in a small quantity of the same fluid: the decoction of them passed through a filtre was examined, and found to contain iron; so that this metal must have been taken up by the vessels or pores in the root.

If to these facts are added those connected with the changes which animal and vegetable substances undergo by the process of putrefaction, we have all the data necessary for forming a rational theory, to guide us in the management and application of manures.

Davy has very satisfactorily shown the cases in which putrefaction or fermentation should be encouraged, and avoided. As a general rule, it may be stated, that when manure consists principally of matter soluble in water, its fermentation or putrefaction should be prevented as much as possible; but on the contrary, when it contains a large proportion of vegetable or animal fibre, such processes become necessary.

To prevent manures from decomposing, he recommends that they should be preserved dry, defended from the contact of the air, and kept as cool as possible. Salt and alcohol, he observes, appear to owe their powers of preserving animal and vegetable substances to their attraction for water, by which they prevent its decomposing action, and likewise to their excluding air. The importance of this latter circumstance he illustrates by the success of M. Appart's method of preserving meat.

By allowing the fermentation of manure to proceed beneath the soil, rather than in the farm-yard, we not only preserve elements which would otherwise be dissipated, but we obtain several incidental advantages; for example, the production of *heat*, which is useful in promoting the germination of the seed. This must be particularly favourable to the wheat crop, in preserving a genial temperature beneath the surface late in autumn, and during winter.

Again:—it is a general principle in chemistry, that in all cases of decomposition, substances combine much more readily at the moment of their disengagement, than after they have been perfectly formed. And in fermentation beneath the soil, the fluid matter produced is applied instantly, even whilst it is warm, to the organs of the plant, and consequently is more likely to be efficient than in manure that has gone through the process, and of which all the principles have already entered into new combinations.

He examines with much attention the various animal and vegetable matters which have been used as manure, and furnishes the farmer with a number of practical remarks on their nature and mode of operation. For these, the reader must refer to the work itself; for my limits will not allow me to enter into the consideration of *rape-cake—malt-dust—linseed-cake—sea-weeds—peat—wood-ashes—fish—bones—hair, woollen rags, and feathers—blood,* &c. &c.; to each of which he assigns peculiar qualities and virtues.

As he regards the due regulation of the fermentative process of the utmost importance, he has furnished some valuable hints for the conduct of the farmer upon this occasion. He considers that a compact marle, or a tenacious clay, offers the best protection against the air; and before the dung is covered over, or, as it were, sealed up, he recommends that it should be dried as much as possible. If at any time it should heat strongly, he advises the farmer to turn it over, and thus cool it by exposure to the air; for the practice sometimes adopted of watering dunghills is inconsistent with just chemical views. It may cool the dung for a short time; but moisture, it will be remembered, is a principal agent in all processes of decomposition.

In cases of the fermentation of dung, there are simple tests by which the rapidity of the process, and consequently the injury done, may be discovered. If, for instance, a thermometer plunged into the mass does not rise above 100°, it may be concluded that there is not much danger of the escape of aëriform matter; but should it exceed this, the dung ought to be immediately spread abroad.

When a piece of paper moistened in muriatic acid, held over the steams arising from a dunghill, gives dense fumes, it is a certain test that the decomposition is going too far; for this indicates that volatile alkali is disengaged.

It may be truly said that, under the hand of Davy, the coldest realities blossomed into poetry: the concluding passage of this lecture certainly sanctions such an opinion, and is highly characteristic of that peculiar genius to which I have before alluded.* A subject less calculated than a heap of manure to call forth a glowing sentiment, can scarcely be imagined.

"The doctrine," says he, "of the proper application of manures from organized substances, offers an illustration of an important part of the economy of nature, and of the happy order in which it is arranged. The death and decay of animal substances tend to resolve organized forms into chemical constituents; and the pernicious effluvia disengaged in the process seem to point out the pro-

* Page 191.

priety of burying them in the soil, where they are fitted to become the food of vegetables. The fermentation and putrefaction of organized substances in the free atmosphere are noxious processes; beneath the surface of the ground they are salutary operations. In this case the food of plants is prepared where it can be used; and that which would offend the senses, and injure the health, if exposed, is converted by gradual processes into forms of beauty and of usefulness; the fetid gas is rendered a constituent of the aroma of the flower, and what might be poison, becomes nourishment to man and animals."

The SEVENTH LECTURE is devoted to the investigation of manures of a mineral origin. He commences the subject by refuting the opinion of Schrader and Braconnot, that the different earthy and saline substances found in plants arise from new arrangements of the elements of air and water, by the agencies of their living organs.

In 1801, he made an experiment on the growth of oats, supplied with a limited quantity of distilled water, in a soil composed of pure carbonate of lime. The soil and the water were placed in a vessel of iron, which was included in a large jar, connected with the free atmosphere by a tube, so curved as to prevent the possibility of any dust, or fluid, or solid matter, from entering into the jar. His object was to ascertain whether any siliceous earth would be formed in the process of vegetation; but the oats grew very feebly, and began to be yellow before any flowers formed. The entire plants were burnt, and their ashes compared with those from an equal number of grains of oat. Less siliceous earth was given by the plants than by the grains; but their ashes yielded much more carbonate of lime.

Numerous other authorities might be quoted to the same effect. Jacquin states that the ashes of Glasswort (*Salsola-Soda*) when it grows in inland situations, afford the vegetable alkali; but when on the sea-shore, the fossile or mineral alkali. Du Hamel also found, that plants which usually grow on the sea-shore, made small progress when planted in soils containing little common salt. The Sunflower, when growing on lands not containing nitre, does not afford that substance; though when watered by its solution, it yields nitre abundantly. De Saussure made plants grow in solutions of different salts; and he ascertained that, in all cases, certain portions of the salts were absorbed by the plant, and found unaltered in their organs.

It may be admitted then as established, that the mineral principles found in plants are derived from the soils in which they vegetate. This fact becomes the foundation of the theory respecting the operation of mineral manure.

Davy observes, that "the only substances which can with propriety be called fossile manures, and which are found unmixed with the remains of any organized beings, are certain alkaline earths, or alkalies, and their combinations." If he intends to limit the term to those bodies only which find their way into the structure of plants, his definition may be correct; but I am inclined to take a much wider view of the subject, and to include all those mineral substances which promote vegetation by modifying the texture of the soil:—but of this hereafter.

Lime, not only from its importance, but from the controversies which it has occasioned, ranks first in the list of mineral manures.

That disputes concerning the uses of lime and its carbonate, should have long existed, and be still continued amongst a class of persons who, whatever may be their practical knowledge, are not acquainted with the composition of the substances about which they differ, is certainly by no means extraordinary. Davy, therefore, very properly introduces the subject, by a description of the nature and qualities of these bodies, and by marking the distinctions between quick-lime and its carbonate.

The substance commonly known by the name of *Limestone* is a compound of lime and carbonic acid, associated generally with other earthy bodies, the nature and proportions of which vary in different species. "When a limestone does not copiously effervesce in acids, and is sufficiently hard to scratch glass, it contains siliceous, and probably aluminous earth. When it is deep brown or red, or strongly coloured of any of the shades of brown or yellow, it contains oxide of iron; when it is not sufficiently hard to scratch glass, but effervesces slowly, and makes the dilute nitric acid in which it effervesces milky, it contains magnesia; and when it is black, and emits a fetid smell if rubbed, it contains coally or bituminous matter."

As the agricultural value of limestone is materially modified by the substances with which it may be associated, their analysis becomes an object of much importance, and the author has accordingly proposed a simple method of effecting it.

Before any opinion can be formed of the manner in which these different ingredients operate, it is necessary that the action of the pure calcareous element as a manure should be thoroughly understood.

In its caustic state, whether used in powder, or dissolved in water, lime is injurious to plants. Davy informs us that he has, in several instances, killed grass by watering it with lime water; but in its combination with carbonic acid, it is an useful ingredient in soils.

When newly-burnt lime is exposed to the atmosphere, it soon falls into powder, from uniting with the moisture of the air; and the same effect is immediately produced by throwing water upon it, when it heats violently, and the water disappears: in this state it is commonly called *slacked* lime: chemists have named it the *hydrat* of lime; and when this hydrat becomes a carbonate, by long exposure to the air, its water is in part expelled, and the carbonic acid takes its place.

Lime, whether freshly burnt, or slacked, acts powerfully on moist fibrous vegetable matters, and forms with them a compost, of which a part is usually soluble in water. By this operation, it renders inert vegetable matter active; and as charcoal and oxygen (the elements of carbonic acid) abound in vegetables, it is itself, at the same time, converted into a carbonate. But limestone simply powdered, marls, or chalks, do not thus act on vegetable matter; and hence the operation of quicklime and mild lime depends on principles altogether different. Quicklime acts on any hard vegetable matter, so as to render it more readily soluble; the mild limes, or carbonates, act only by improving the texture of the soil, or by supplying a due proportion of calcareous matter: thus almost all soils which do not effervesce with acids, are improved by mild lime and sand, more than clays. I apprehend that it is upon this principle the application of shelly sand proves beneficial in Cornwall, although I have ascertained that, on some occasions, its value depends upon its chemical action upon mineral bodies in the soil.

Soils abounding in soluble vegetable manures are injured by quicklime, as it tends to decompose their soluble matters, or to form with them compounds less soluble than the pure vegetable substance. With animal manures, it is equally exceptionable, unless indeed they be too rich, or it becomes necessary to prevent noxious effluvia: for since it decomposes them, it destroys their efficacy, and tends to render the extractive matter insoluble.

The limestones containing alumina and silex are less fitted for the purposes of manure than pure limestones; but the lime formed from them has no noxious quality. Such stones are less efficacious, merely because they furnish a smaller quantity of quicklime. Those, however, that contain magnesia, if indiscreetly used, may be very detrimental.

It had been long known to farmers in the neighbourhood of Doncaster, that lime made from a certain limestone, when applied to the land, often injured the crops considerably. Mr. Tennant discovered that this limestone contained magnesia; and on mixing some calcined magnesia with soil, in which he sowed different seeds, he found that they either died, or very imperfectly vegetated; and with great justice and ingenuity, he referred the bad effects of the peculiar limestone to the magnesian earth it contained. In prosecuting the enquiry, Davy however ascertained that there were cases in which this magnesian lime was used with good effect,—in small quantities, for example, on rich land: and during his chemical consideration of the question, he was led to the following satisfactory solution.

"Magnesia has a much weaker attraction for carbonic acid than lime, and will remain in the state of caustic or calcined magnesia for many months, though exposed to the air; and as long as any caustic lime remains, the magnesia cannot be combined with carbonic acid, for lime instantly attracts carbonic acid from magnesia. When therefore a magnesian limestone is burnt, the magnesia is deprived of its carbonic acid much sooner than the lime, and in this state it is a poison to plants. That more magnesian lime may be used upon rich soils,*

* These facts have been confirmed by agriculturists, who could not possibly have had any favourite theory to support. Dr. Fenwick tells us, (Essays on Calcareous Manures, p. 11. 1798,) that in the county of Durham, the farmers always distinguish between *hot* and *mild* limes. They never apply the former to exhausted lands, or to any soil that has been long under a course of tillage, unless it be very deep and rich. In peaty soils, and in new, sour, and wild lands, the *hot* limes, on the contrary, are preferred to the *mild* ones. Dr. Fenwick made some experiments to ascertain the cause of the differences between these varieties of lime; and

seems to be owing to the circumstance, that the decomposition of the manure in them supplies carbonic acid, and thus converts it into a mild carbonate. Besides being used in the forms of lime and carbonate of lime, calcareous matter is applied for the purposes of agriculture in other combinations. The principal body of this kind is *gypsum*, or sulphate of lime; respecting the uses and operation of which very discordant opinions have been formed.

Its beneficial operation has been referred to two causes, viz. to its power of attracting moisture from the air, or to its assisting the putrefaction of animal substances; but Davy has shown by experiments that neither of these theories can be supported by facts.

The most extraordinary circumstance perhaps connected with the history of this mineral manure, is the very opposite opinions which have been formed respecting its value. In this country, although there are various testimonies in its favour, it has never been employed with the signal success which marked its adoption in America, and which was so palpable and extraordinary as at once to have ensured its universal introduction.

I was some years since assured by Mr. Maclure of Philadelphia, that whenever any doubt or hesita-

though he failed to discover that by analysis which Mr. Tennant subsequently ascertained, he nevertheless arrived at a just conclusion by simple observation; and was led to believe, that "what farmers term *hot* limes, are such as re-absorb their fixed air more slowly, and therefore continue longer to exert the peculiar action of quicklime."

tion betrayed itself with respect to its fertilizing agency, it was only necessary to sprinkle a small quantity in a meadow, to satisfy the most sceptical; and that this was usually done in the form of letters or characters, which in a short time became so much more luxuriant than the surrounding grass, as to be visible at a considerable distance. It is, I understand, chiefly applied to grass lands as a *top-dressing;* and the American farmers* explain its operation upon its solubility in water, and its consequent absorption by the roots of the grass. Davy, in examining the ashes of sainfoin, clover, and rye-grass, which had grown in soils manured by gypsum, found considerable quantities of that substance; and he thinks it probable that it was intimately connected with their woody fibre. He attempts to explain the reason why the application of gypsum is not generally efficacious, by supposing that most of the cultivated soils may already contain it in

* When this substance was first introduced into America, which is nearly forty years since, it was imported from the quarries of Montmartre, and in such request was it, that a bushel of wheat was usually given for the same measure of gypsum: it is now, I believe, obtained from Nova Scotia; I have not heard that it has been found within the States. It may perhaps serve to convey some idea of the extent to which it has been applied, when I state, that Mr. Maclure assured me that not less than three hundred vessels are constantly employed in the traffic, and that in Philadelphia twenty merchants, at least, are engaged in supplying the demand for it. Its efficacy appears to be considerably increased by applying it in a minute state of division; and a want of attention to this circumstance may possibly have been one of the causes which have rendered its advantages less conspicuous in England. In America, three or four hundred mills, of a peculiar construction, have been erected in different parts for the purpose of grinding it.

sufficient quantities for the use of the grasses. I strongly suspect, however, that it will be hereafter discovered to depend upon the nature of the soil in its hygrometric relations. From the facts already recorded, it would appear that it never answers near the sea, nor in wet lands. In consequence of its solubility, it is enabled to penetrate and pervade the whole vegetable structure; and the experiments of Davy have proved its presence in the ashes of plants exposed to its operation, and have rendered it probable that it enters into union with their woody fibre, by which the density of their textures will be increased, and consequently the evaporation from their leaves diminished; I am from such considerations induced to think that gypsum does not act by effecting any chemical change in the soil, but solely by diminishing the plant's evaporation. This idea seems to be borne out by the evidence furnished by the different circumstances attending the operation of this manure: we find, for example, that succulent vegetables, planted on dry soils, are those which are principally benefited by its application, and that the various grasses so manured retain their verdure, even in the dryest season and on the most arid lands; at the same time, we find that these crops, especially clover, acquire a proportionate increase in the density of their fibres, that is to say, that they become much more rank and stubborn, and often to such a degree does this take place, that in America, where its effects are best understood, sheep not uncommonly refuse to feed upon them. Upon the same principle we find that, under circumstances or in situations where the evaporation of a plant is provided for by a constant supply of moisture, the effects of gypsum cease to be apparent.

Davy hints at a process by which gypsum may be formed in a soil containing sulphate of iron, by the action of calcareous manure,* and which was first pointed out by Dr. Pearson. I can confirm this statement by the results of experiments I formerly made in Cornwall, where soil containing this salt of iron had been manured by shelly sand.

In pursuing his enquiry into the efficacy of mineral manure, Davy proceeds to investigate the efficacy of the fixed alkalies, and observes that their general tendency is to give solubility to vegetable matters, and in this way to render carbonaceous and other substances capable of being taken up by the tubes in the radical fibres of plants. The vegetable alkali has likewise a strong attraction for water, and even in small quantities may tend to give a due degree of moisture to the soil, or to other manures.

He considers that pure salt may act, like gypsum, phosphate of lime, and the alkalies, by entering into the composition of the plant. Upon the subject of salt, however, his remarks are very meagre and unsatisfactory: at the time he composed his lecture, the subject had not excited that public attention which the writings of Mr. Parkes, Sir Thomas Bernard, and others, have since awakened.

* Gypsum is readily produced by the admixture of decomposing pyrites and calcareous matter: in proof of which the Mineralogist can produce specimens of oyster shells studded with crystals of selenite from Shotover; and alum from the *aluminous shale* at the Hurlet Mine near Glasgow.

Had our philosopher undertaken the agricultural survey of Cornwall, his lecture on mineral manure must have been very considerably extended. He would have learnt that various rocks reduced to small fragments, are commonly applied as dressing; he would have explained the cause of the fertility so generally associated with hornblende rocks;—he would have speculated upon the influence of iron in giving fruitfulness; and above all, he would have taught the agriculturist the scientific use of calcareous sand, by pointing out the description of lands which are most likely to be benefited by its application.

The EIGHTH LECTURE concludes the subject of the chemistry of Agriculture, by establishing the theory of the operation of burning lands. He considers the process to be useful in rendering the soil less compact, and less tenacious and retentive of moisture; and that, when properly applied, it is capable of converting a matter that was stiff, damp, and cold, into one powdery, dry, and warm, and much more proper as a bed for vegetable life. He states the great objection made by speculative chemists to paring and burning, to be the unavoidable destruction of vegetable and animal matter, or the manure of the soil; but he considers that, in those cases in which the texture of its earthy ingredients is permanently improved, there is more than a compensation for so temporary a disadvantage; and that in some soils, where there is an excess of inert vegetable matter, the destruction of it must be beneficial, and that the carbonaceous matter remaining in the ashes may be more useful to the crop than the vegetable fibre from which it was produced.

In this view of the subject it is evident, that all poor siliceous sands must be injured by the operation; "and here," says Davy, "practice is found to accord with theory. Mr. Arthur Young, in his Essay on Manures, states, 'that he found burning injure sand;' and the operation is never performed by good agriculturists upon siliceous sandy soils, after they have been once brought into cultivation. An intelligent farmer in Mount's Bay told me, that he had pared and burned a small field several years ago, which he had not been able to bring again into good condition. I examined the spot,—the grass was very poor and scanty, and the soil an arid siliceous sand." *Irrigation*, or *watering land*, is a practice, he observes, which at first view appears the reverse of torrefaction; and, in general, the operation of water in nature is to bring earthy substances into an extreme state of division. But in the artificial watering of meadows, the beneficial effects may depend upon many different causes, some chemical, some mechanical. It may act as a simple supply of moisture to the roots, or it may carry into the soil foreign matter, or diffuse that which exists in it more equally through its substance.

He concludes with some valuable scientific observations upon the process of *fallowing*, by which he attempts to correct the prejudices which have existed with regard to its benefits. He points out, on the other hand, the great advantages of the convertible system of husbandry, by which the whole of the manure is employed; and those parts of it which are not fitted for one crop, remain as nourishment for another. These views he illustrates by a reference to the course of crops adopted by Mr. Coke, in which "the turnip is the first in the order of succession; and this crop is manured with recent dung, which immediately affords sufficient soluble matter for its nourishment; and the heat produced in fermentation assists the germination of the seed and the growth of the plant. After turnips, barley with grass seeds is sown; and the land having been little exhausted by the turnip crop, affords the soluble parts of the decomposing manure to the grain. The grasses, rye-grass, and clover remain, which derive a small part only of their organized matter from the soil, and probably consume the gypsum in the manure which would be useless to other crops: these plants likewise, by their large system of leaves, absorb a considerable quantity of nourishment from the atmosphere; and when ploughed in at the end of two years, the decay of their roots and leaves affords manure for the wheat crop; and at this period of the course, the woody fibre of the farm-yard manure, which contains the phosphate of lime and the other difficultly soluble parts, is broken down; and as soon as the most exhausting crop is taken, recent manure is again supplied."

At the end of his system is added an Appendix, containing "An Account of the results of Experiments on the produce and nutritive qualities of the Grasses and other plants used as the food of animals; instituted by John Duke of Bedford." But as these experiments do not admit either of abridgement or analysis, the reader must refer to the original source for information.

I shall conclude this long, and, I fear, somewhat tedious review, with the animated appeal so earnestly addressed by the illustrious author to the philosophical readers of his work.

"I trust that the enquiry will be pursued by others; and that in proportion as chemical philosophy advances towards perfection, it will afford new aids to agriculture: there are sufficient motives connected both with pleasure and profit, to encourage ingenious men to pursue this new path of investigation. Science cannot long be despised by any persons as the mere speculation of theorists, but must soon be considered by all ranks of men in its true point of view, as the refinement of common sense guided by experience, gradually substituting sound and rational principles for vague popular prejudices.

"The soil offers inexhaustible resources, which, when properly appreciated and employed, must increase our wealth, our population, and our physical strength.

"We possess advantages in the use of machinery, and the division of labour, belonging to no other nation. And the same energy of character, the same extent of resources, which have always distinguished the people of the British Islands, and made them excel in arms, commerce, letters, and philosophy, apply with the happiest effects to the improvement of the cultivation of the earth. Nothing is impossible to labour, aided by ingenuity. The true objects of the agriculturist are likewise those of the patriot. Men value most what they have gained with effort; a just confidence in their own powers results from success; they love their country better, because they have seen it improved by their own talents and industry; and they identify with their interests the existence of those institutions which have afforded them security, independence, and the multiplied enjoyments of civilized life."

END OF THE FIRST VOLUME.

LONDON:
PRINTED BY SAMUEL BENTLEY,
Dorset Street, Fleet Street.

Printed in the United States
By Bookmasters